Web Marketing for the Music Business

Web Marketing for the Music Business

Tom Hutchison

ELSEVIER

AMSTERDAM • BOSTON • HEIDELBERG • LONDON
NEW YORK • OXFORD • PARIS • SAN DIEGO
SAN FRANCISCO • SINGAPORE • SYDNEY • TOKYO

Focal Press is an imprint of Elsevier

Focal
Press

Focal Press is an imprint of Elsevier
30 Corporate Drive, Suite 400, Burlington, MA 01803, USA
Linacre House, Jordan Hill, Oxford OX2 8DP, UK

Library of Congress Cataloging-in-Publication Data
Hutchison, Thomas W. (Thomas William)
  Web marketing for the music business/Tom Hutchison.
    p. cm.
  Includes bibliographical references and index.
  ISBN 978-0-240-81044-7 (pbk. : alk.paper) 1. Music trade. 2. Internet marketing. I. Title.
ML3790.H986 2008
780.68'8–dc22

                                                                      2008020851

British Library Cataloguing-in-Publication Data
A catalogue record for this book is available from the British Library.

ISBN: 978-0-240-81044-7

For information on all Focal Press publications
visit our website at www.books.elsevier.com

08 09 10 11 12 5 4 3 2 1

Printed in the United States of America

Table of Contents

CHAPTER 4 • Domains and Hosts: Nuts and Bolts

CHAPTER 5 • Creating the Web Site

CHAPTER 6 • HTML and Scripts

CHAPTER 7 • Optimizing and Monitoring Your Web Site to Increase Visitation

CHAPTER 8 • Audio and Video for Your Web Site

CHAPTER 9 • E-commerce: Product Ordering and Fulfillment

CHAPTER 10 • Finding Your Online Market

CHAPTER 11 • Successful Promotion on the Web

CHAPTER 12 • Social Networking Sites

CHAPTER 13 • Internet Obstacles

CHAPTER 14 • Mobile Media

CHAPTER 15 • Mobile Music Marketing

Online Resources for the Music Business

Acknowledgments

There are many people I would like to thank for their support and participation in the creation of this book. I would like to thank my colleague Paul Allen, for convincing me to write this book based on a class we developed and teach, and colleague Amy Macy, who has been sharing her web discoveries with me. Thanks to colleague Storm Gloor, who initially reviewed the project. Thanks to my editor Catharine Steers and editorial assistant David Bowers of Focal Press. Thank you to MTSU alumni John Haring of Nashville Independent Music, who has been of great assistance as technical editor and footnote author, and Stephanie Orr-Buttrey of Country Wired.com, who has contributed her expertise and access to artists she represents. Thanks to all the musicians who allowed me to use them as examples throughout the book: Bill "Sauce Boss" Wharton, Richard and Andy Karg, Grant Peeples, Dick Hosford, and Inga Swearingen.

I would also like to thank the employees of many companies who were gracious enough to help me obtain permission to use their images in the book, including Gina Wilkinson of imeem, Lindsay Marlow of PassAlongNetworks, Ryan Jensen of The Ezine Directory, Grant Leavins of Yahoo!, Andrew Lu of FastStone Soft, several wonderful people at MySpace, and others whose works are featured throughout the book. Thanks to all of you.

Thank you to my two student research assistants, Jennifer Harbin, who trained me after I trained her on using social networks, and Brent Powelson, who was a reader for several chapters. And most of all, I'd like to thank my wife, Lynne, who assisted when she could but always filled me with inspiration to write—and finish—this book.

Preface

The introduction of the compact disc in 1983 revived the recorded music industry, which had been in a slump since 1979. This revolutionary new format for disseminating recorded music proved to be a boon for companies and consumers alike. Certain factors contributed to the overwhelming financial success of the format. The CD format is standardized, so there are no compatibility issues, and the advantages to the consumer were obvious. CDs were initiated by the record labels and partnering hardware companies rather than outside businesses. Most important for the record labels, the music that consumers currently owned was not transferable—consumers needed to pay to replace their vinyl and cassette collections.

The industry experienced a similar revolution in the late 1990s, but this time the record labels chose not to participate. Indeed, they attempted to thwart the coming age of digital downloads—a move that contributed to an industry downturn that has been occurring since 2000.

This should be a time of great opportunity in the music industry. As Chris Anderson pointed out in his book *The Long Tail*, the costs of creating musical works, the cost of dissemination, and the costs of marketing have all been reduced by computers and Internet access. There is more music being created and recorded today than at any time in history. There is more music being consumed today, as fans carry their entire music collections with them on portable devices. But for the record business, the transition has proved difficult. The record labels thought they were sailing toward a future of continued growth in CD sales. Then they were hit by the perfect storm.

Consumers prefer cherry-picking songs, buying the singles they like without being forced to buy an entire CD. But with the demise of the physical single (no longer found in any format), consumers had been forced to turn to albums to get the songs they like. My research at the time indicated that consumers were reluctant to "pay 18 dollars to get the one or two songs" they wanted. The adoption of digital downloads would offer the labels the ability to revive the "single" in a format that yields a profit at a reasonable price, while satisfying the customer preference for cherry-picking.

And even in the absence of an initiative on the part of the industry to provide their music in a digital downloading format, consumers demonstrated their readiness to embrace the MP3 format and move their music collections from the CD tower to the hard drive. First came illegal peer-to-peer file sharing, facilitated by MP3.com and followed by Napster. There were no legal alternatives for the consumer to download those coveted songs. Concerned for the security of their intellectual property, record labels were reluctant to license music to downloading services. Then Steve Jobs created a plan to introduce a closed system that would protect the copyright holder while allowing consumers to cherry-pick and download tracks to their portable iPods. A few years later, with iTunes holding over 70% of the market, the labels became concerned that Jobs was wielding too much control over the industry. They began looking for other ways to sell music.

Numerous business models are in the beta-testing stage at the time of this writing. Those include subscription music services, copy-protected and non-copy-protected downloads, advertiser-sponsored music, streaming music services, and probably a few others being discussed behind closed doors. Despite the confusion in how to best profit from music so that the industry can continue to develop new talent, there is an abundance of opportunity for reviving the industry.

The one certainty is that the compact disc is in a phase of decline and is unlikely to experience a revival. For all its attributes—including fidelity, portability, durability, and the cool, shiny look—the CD has its share of negatives. Compared to digital downloads, the CD is inconvenient, bulky, costly, and a burden on the environment.

According to the Environmental Protection Agency, about 100,000 CDs and other optical discs become waste each month. It's difficult to determine the total environmental impact of using CDs to disseminate music, movies, and computer software. However, one study estimates an impact of one kilogram of carbon dioxide equivalents for each music CD that is produced, packaged, and delivered (about one-third the amount of a paperback book). Another study cites a one-kilogram difference between physical CD distribution and the electronic distribution equivalent—half for manufacture of the disc and half for distribution. www.Earth2tech.com asserts that the materials and additives in CDs and jewel cases include polyvinyl chloride (PVC), a polycarbonate plastic, an aluminum layer over the plastic, and resin. Compact discs are not biodegradable, and when incinerated they release toxins into the atmosphere. Earth2tech states, "Publishers and vendors in these markets [CD and DVD] can and should lessen their impact on the environment by moving to 'green' electronic distribution." Perhaps more artists should consider adopting reusable jump drives to store and sell music at live venues and other places where physical product is still important. And otherwise, they should support licensed downloading services for the electronic dissemination of their works. Until then, these artists should encourage the recycling of optical discs.

This book offers numerous tips on how to move from CDs to new, virtual music marketing, promotion, discovery, and acquisition. In the recording industry, we have entered a brave new world—it's time for the music business to lead instead of follow.

—Tom Hutchison

CHAPTER 1

Development of Music Marketing on the Internet

BACKGROUND ON THE INTERNET

The *Internet* is described as a global network connecting millions of computers. Unlike online services, which are centrally controlled, the Internet is decentralized by design. Each Internet computer, called a host, is independent. The Internet got its start as several universities and government organizations saw the benefit in connecting mainframe computers together to share information. The military saw its use for communication in case the country was ever under attack. Little did these innovators realize that in 2007, a blockbuster Bruce Willis movie would feature the country under attack *of* just such an infrastructure.

Electronic mail was introduced in the early 1970s along with communication protocols still in use today: *transmission control protocol/Internet protocol* (TCP/IP) and *file transfer protocol* (FTP). The domain name system was introduced in 1985 with the extensions .com, .org, .net, .gov, .mil, and .edu. Then, in 1989, Tim Berners-Lee of CERN (European Laboratory for Particle Physics) developed a new technique for distributing information on the Internet he called the World Wide Web. Based on hypertext, the Web permits the user to connect from one document to another at different sites on the Internet via *hyperlinks* (specially programmed words, phrases, buttons, or graphics). Unlike other Internet protocols, such as FTP and e-mail, the Web is accessible through a graphical user interface. This brought the development of web browsers that could decipher coding in documents and display them according to a set of instructions called hypertext markup language. Meanwhile, the first efforts to index the contents of the Internet were introduced in the early 1990s.

In 1995, several online services (America Online, CompuServe, and Prodigy) began providing their software to computer manufacturers to include on new computer hard drives.[1] This propelled Internet use as new computer owners took advantage of the free trial period offered by several of these services. By 1996, 43.2 million (44%) U.S. households owned a personal computer, and 14 million of them were online.

In 1999, college student Shawn Fanning introduced *Napster*, a program that allowed users to swap music files over the Internet. This began the downfall of the recorded music industry as it had existed until that time. By 2001, the courts had ruled that Napster was in violation of copyright laws and forced it to shut down its file-sharing operations. New file-sharing services sprang up to fill the void, including Aimster, Grokster, Kazaa, Limewire, and BitTorrent. By January of 2002, 58.5% of the U.S. population was using the Internet. By 2003, it was estimated that the illegal sharing of music files had grown to about 2.6 billion files per month.

Web 2.0

In 2001, the "dot-com" bubble started to burst as traditional companies began to pull back on their Internet advertising budgets. Massive layoffs followed as content providers sought less expensive ways to provide content for Internet users. This brought about the development of *Web 2.0*, a system in which users are the content providers and the Web 2.0 companies just provide the platform. Media writer and analyst Tim O'Reilly, the founder and CEO of O'Reilly Media, coined the term *Web 2.0*. Web 2.0 is basically the current generation of popular web applications that are democratic in nature, put the user in control, and rely on the aggregate wisdom of the masses to provide content and feedback. One of the early examples is Wikipedia, launched in early 2001 as "harnessing the wisdom of crowds to build an online encyclopedia" *(Information Week)*; it went live in 2005.

> The concept of "Web 2.0" began with a conference brainstorming session between O'Reilly and MediaLive International. Dale Dougherty, web pioneer and O'Reilly VP, noted that far from having "crashed," the web was more important than ever, with exciting new applications and sites popping up with surprising regularity. What's more, the companies that had survived the collapse seemed to have some things in common. Could it be that the dot-com collapse marked some kind of turning point for the web, such that a call to action such as "Web 2.0" might make sense? We agreed that it did, and so the Web 2.0 Conference was born.
>
> **Tim O'Reilly, www.oreillynet.com**

---

[1] These companies, as well as GEnie and other smaller ones, developed dial-up systems as early as the late 1960s, but 1995 saw the first widely used WWW access and bundling.

MySpace was founded in 2003, which allowed casual users to build a web page for themselves and build social networks online. Flickr launched in early 2004, giving users the opportunity to store and share photos online. Yahoo bought Fickr in 2005. Other Web 2.0 innovations include Craigslist, Ebay, Digg, Facebook, YouTube, Second Life, etc. In 2006, *Time Magazine* awarded its person of the year award to "you—the creators on Web 2.0." *Time* describes Web 2.0 as "a tool for bringing together the small contributions of millions of people and making them matter." By 2006, there were also more than 92 million web sites online.

Another innovation that has breathed life into Web 2.0 is *Google AdSense*. Introduced in 2005, AdSense provides a way for web site hobbyists to monetize their labor-of-love web pages. The millions of noncommercial web sites that were set up and run for personal interest now had a means to make money from their visitors by running ads. Many of these sites were simply advice, how-to, or other informational sites that did not necessarily sell products or charge money for the information they provide. So, with little setup and at no cost, these mom-and-pop web sites were able to sign up on Google to participate in AdSense and start allowing Google to run topic-related ads on these sites. Site owners are paid a small sum each time a visitor "clicks through" one of the ads to visit the advertiser's web site. The advertisers participate through the other side of the equation, Google AdWords.

Table 1.1	Some of the Top Web 2.0 Sites	
Digg: Suggest stories for the votes of a wider readership.	**Craigslist:** Online classified ads for free.	**Wikipedia:** An online encyclopedia with users contributing content.
Flickr: Post and share digital photos online.	**Slide.com:** Create slide show and share or embed into your social networking page.	**YouTube:** Upload videos and share or embed into your social networking page.
MySpace: An online community with more than 70 million members, most notably teenagers and bands.	**Meebo:** Access your buddy list and IM all you want from its web site.	**Second Life:** A virtual world where users can create an alternate life and interact with others.
Facebook: A social-networking site particularly popular among college students.	**Twitter.com:** A text messaging service that lets people send notes to groups.	**imeem:** Share music, video, photos, comments, and blogs with friends.
Bebo: Share photos and blogs, draw on your own and other member's White Boards, and find school and college friends.	**Amazon marketplace:** E-commerce for small vendors.	**LastFM:** Create your personal Internet radio station by rating songs.

(Continued)

Table 1.1	Some of the Top Web 2.0 Sites—cont'd	
Pandora: Creates a radio station based on collaborative filtering based on your musical preferences.	**Eyspot:** Upload your video and use the site's tools to edit it and publish it on other sites.	**LinkedIn:** Social networking site for career advancement.
NetFlix: Uses collaborative filtering to allow its members to rate movies and receive recommendations.	**Blogger.com:** A site owned by Google that makes it easy to start and maintain a blog.	**Skype:** Audio and video communication via the Internet.
eBay: An online auction where eBay members sell items to other members.	**Dodgeball:** Lets users alert their friends to where they are via cell phone messages.	**Google Earth:** Create mashups of aerial photographs overlaid with photos and descriptions.

MUSIC AND THE INTERNET

When the Internet was in its infancy in the late 1980s, only a few audiophiles and tech-savvy musicians were online, communicating through the few rudimentary online services such as *CompuServe* and General Electric's *Genie*. The author of this book was one of those early online music pioneers, as a systems operator on CompuServe's MIDI forum. Modem baud rates were such back then that even small text files took a while to download, and the online services were charging by the minute. One of the most progressive music exchanges occurring at the time were in the form of MIDI data files,[2] small enough to move through the dialup systems. The MIDI forum had three sections: (1) the message section, where members could post messages and exchange ideas, (2) the libraries section, which hosted the MIDI library of songs, and (3) the conference room, where members held live chats, sometimes featuring a well-known artist or producer. As musicians started to exchange ideas via MIDI files over the Internet, the first online musical collaborations were created. A musician would start a song, record the data in MIDI, upload the file to the forum or send it to a particular forum member, and then allow that member to modify the file, thus modifying the composition. The forum was also a place to exchange ideas, and some members published articles about music and the music business. The forums

---

[2] MIDI files are the digital equivalent of a player piano scroll or music box disc. Bits of information instruct musical notes on a specialized musical instrument to turn on and off at particular points in the song. For MIDI, these are simple, small files that activate and direct the more complicated musical sound modules to create complex sounds that emulate a host of musical instruments.

were not consumer oriented at first. Then in the early 1990s, Geffen Records decided to have an online presence for fans and opened up a Geffen forum on CompuServe.

GEFFEN AND COMPUSERVE

Geffen provides CompuServe and Internet users access to graphics, sound clips, tour schedules and artist biographies. Geffen artists interact with users with on-line messages and conferences. Geffen regularly runs contests offering prizes to promote on-line use. Geffen employees surf the Internet, CompuServe and other on-line services in the course of their work, including the pursuit of new talent for the label.

ComputerWorld Honors, 1995

By 1994, Geffen Records was trying new ways to reach consumers through the Internet with its debut of an Aerosmith song online through its CompuServe forum (EW.com, 1994). For one week, they offered an outtake of the song "Head First" from the band's *Get a Grip* album. Early Internet adopters were able to download the song, which took up to several hours; the band waived their royalties while CompuServe suspended their usual $9.60 hourly charge. Geffen's Jim Griffin commented at the time "We're not saying this is how you'll get your music in the future. But we did want to try it out." Shortly after that, Geffen was one of the first labels to create its own web site. One way to look at old web pages from history is through the "Wayback Machine" at www.archive. org/web/web.php. Some of the original Geffen Records pages and assets are archived at http://web.archive.org/web/*/geffen.com/*.

In 1995, Polygram Records introduced a prototype record label web site for consumers at the 1995 National Association of Recording Merchandisers (NARM) convention in San Diego. The label touted to skeptical music retailers that the site was not designed to siphon off retail business but to guide consumers to record stores that carried the products featured on the site. Soon after that, scores of record labels and artists began setting up web sites and procuring their domain names while tiptoeing around the issue of cutting into retail store sales.

In his 2001 book *The Musician's Internet*, writer Peter Spellman presented this comment that he wrote in 1995:

Online distribution—actually delivering that album order not through the mail but as digital bits down telephone or cable wires—is still pretty much in the "wouldn't it be neat?" stages...But the fact remains that the entire history of recorded music will soon be available for instant downloading. The technology to do this already exists, though availability to the public is still a few years away.

Peter Spellman, *The Musician's Internet*, 2001

MP3 Format

By 1997, the first viable audio compression format, *MP3*, was being introduced to consumers through several computer audio players such as the popular WinAmp.[3] Consumers were able to take audio files from their CDs or master tapes, compress the files, and e-mail them to other Internet users. In 1998, the web site MP3.com launched after noticing an abundance of web traffic using the search term *MP3* to look for free music files. The site offered a way for people to post their favorite (or original) music files to share with others and was, at that time, strictly an advertiser-supported project. "At its peak, MP3.com delivered over 4 million MP3 formatted audio files per day to over 800,000 unique users on a customer base of 25 million registered users" (Wikipedia). In 2000, MP3.com launched a new service, MyMP3.com, which allowed registered users to upload a copy of music they already owned and then stream it to the computer of their choice. The company saw this as a legitimate way to honor copyright protection, but the Recording Industry Association of America (RIAA) and the labels did not agree, and sued MP3.com. The company settled with the labels and the service was discontinued, but by then Napster and other illegal *peer-to-peer* (P2P) file sharing networks were in full swing.

FIGURE 1.1
Diamond MP3 Player from Rio. (Courtesy Diamond Multimedia.)

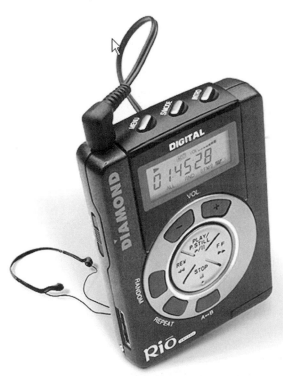

Also in 1997, the Rio was introduced as the first mass market MP3 player. The RIAA immediately sued the company, Diamond, claiming the device allowed for copyright infringement and storage of illegal copies of songs. The RIAA claimed the Rio was *made* for illegal pirating. The courts disagreed, and the Rio hit the market. In upholding this decision, the appeals court said the Audio Home Recording Act, which the RIAA had cited in its suit, prohibits devices that make copies from digital music recordings. The court said the Rio does not make copies but simply stores files it gets from computers (*Salon*, June 16, 1999). But the original Rio players were limited: storage capacity topped out at 64 megabytes (MB) but most had only 32 MB, and data transfer via the parallel or USB 1.0 port was very slow. Several companies created audio software to allow for MP3 encoding and playback facilities on the computer, both for the MacIntosh and the personal computer (PC) platforms.

The original players were more of a novelty than a practicality: "for a mere $200, [you can]

---

[3] WinPlay3 was the first mp3 player for PC's. It came out in 1995. WinAmp was second, in 1997.

listen to 60 minutes of CD-quality music in the MP3 audio format" (Brown, 1998). Users had to return to their computers and reload the player to listen to more songs or use removable media such as SmartMedia cards to hold additional music. By 1999, portable MP3 players from a number of manufacturers were popular, and other software developers, such as Liquid Audio and a2b, scrambled to develop audio compression formats that were more secure and would appeal to piracy-conscious record labels. In the fall of 1999, Napster was in full swing, and MP3 players were popular with college students. The RIAA began contacting universities and targeting college students, who were the most likely copyright violators. Access to high-speed Internet on campuses was fueling massive peer-to-peer music file sharing. The RIAA sent notices to more than 300 universities warning them that students were hosting illegal MP3 files on campus servers. Janelle Brown, writer for *Salon*.com, wrote in November 1999 "there are probably millions of illegal MP3 files and music traders online—pity the poor fool whose job it is to track them all down" (Brown, 1999a).

By 2001, Brown was writing that the promise of a digital revolution that takes power away from the major conglomerates and puts it in the hands of the creative producers was all but dead. "Innovation is being sledge-hammered out of existence by legal threats and buyouts" (Brown, 2001). Vivendi Universal purchased MP3.com, BMG bought Myplay.com, the RIAA had defeated Napster and gone after several other services. Consumer frustration was growing because the major labels were not providing a legal, commercial alternative to illegal downloading at a time when the consumer was more than ready to convert from CDs to digital tracks stored on a computer system. "Napster had single-handedly turned millions of consumers on to the world of MP3s" (Brown, 2001). Many of the startups that avoided legal action were having a difficult time with profitability because the labels were unwilling to license music for legal downloads. The major labels were creating online distribution services of their own, but they failed to capture the commercial market with their cobbled-together group of startups such as MusicNet and Duet. They had yet to hit on a formula for success in making a dent in the illegal P2P file trading, until Steve Jobs of Apple convinced them he had the answer.

ITUNES STORY

In January of 2001, Apple introduced the iTunes jukebox software at Macworld. Unlike the predecessors, iTunes was simple, with most of the screen dedicated to a browser for finding music (www.ipodobserver.com), but the initial release was only for Mac users, with the promise of a Windows version to follow. iTunes included support for creating mixes, burning CDs, and downloading to popular MP3 players, including the Rio. Unfortunately, the current crop of MP3 players did not offer a Mac user-friendly interface. Steve Jobs ordered the development of an MP3 player that would work seamlessly with the iTunes system. Part two of that plan was to create an online music downloading store. Thomas Hormby

FIGURE 1.2
iTunes computer interface. (Courtesy of Apple, Inc.)

wrote, "Apple was not the first to create such a store, but it was the first not to fail spectacularly. The most notable pre-Apple music store was Pressplay, which was a joint venture between major record labels" (Hormby, 2007). Unlike the other legal downloading services that used a cacophony of policy restrictions (limited number of downloads and steams per month), the iTunes policy was simple: 99 cents per download, up to three copies. It was an instant success, quickly capturing 76% of the hardware market and 82% of the legal music download market (Silverstein, 2006).

In 2007, the record labels sought a price restructuring of iTunes but were unable to persuade Jobs, who believed the current pricing strategy of 99 cents was crucial for the continued success of the service. According to an article in *Business Week*, "Universal Music Group chief Doug Morris and his fellow music executives realized they had ceded too much control to Jobs" (*Business Week*, 2007). Morris refused to re-sign a multiyear contract with iTunes and instead opted for a month-to-month agreement while nurturing alternative retailing options, including a new service called Total Music. One of the big question marks about this new business model is that it does not offer anything that has not already been tried—and failed. The plan does call for hardware makers and cell phone makers to absorb the cost of a monthly subscription service—something consumers themselves have so far rejected. Consumers do not seem to want to "rent" music by the month; they prefer to own the music. Consumers are concerned about losing access to their music library if they leave the service or if the service increases their fees. The existing subscription services have only a few million customers, whereas iTunes has hundreds of millions of users.

This new model calls for hardware makers to add the $5 monthly subscription cost (for 18 months) to the price of the hardware, increasing the retail price by $90, so ultimately, the consumer is paying for the subscription service. Meanwhile, Amazon has launched an MP3 retail store and is in the process of licensing music from the major labels. Wal-Mart and Best Buy have also made an entrance into the music download market. Not one of these companies is giving the record labels a new pricing strategy, which they were originally seeking from Apple. And in spring 2008, iTunes moved past Wal-Mart to become the leading music retailer in the United States.

FIGURE 1.3
Third-generation iPods. (Courtesy of Apple, Inc.)

CURRENT ONLINE ENVIRONMENT

The development of social networking sites dedicated to discovering and sharing new music brings hope that the millions of unsigned musicians who have

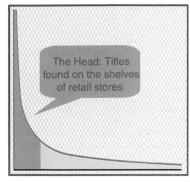

FIGURE 1.4
The head of the
long tail.

FIGURE 1.5
The tail of the
long tail.

recently acquired the technology to create high-quality recordings will now be able to achieve a moderate level of success, creating a middle class of professional entertainers that has been missing in the previous decades. A "record deal" with a major record label may no longer be necessary to enjoy the success of having a large fan base to purchase recorded music, concert tickets, and artist memorabilia. Several chapters in this book are dedicated to outlining how to make this happen.

Chris Anderson, in his book *The Long Tail*, describes how new Internet applications, including *collaborative filtering*, are moving the market away from the "head" consisting of the top-selling artists, writers, movies, and so forth and moving members of the market "down the tail" by allowing them to discover new, less popular products easily that would have, under previous circumstances, gone unnoticed in the marketplace.

In other words, the self-promoted artist and those on small independent labels now have several outlets through which they can promote and sell music to a wider market than in the past, basically lowering the barrier of entry into the music business. Not only are these small enterprises able to promote and distribute music at a fraction of the previous cost, but the newer, sophisticated techniques for music discovery via search engines, collaborative-filtering software, and music networking sites make it easier for music fans to discover and find some of these lesser-known artists that appeal to them.

GLOSSARY

Collaborative filtering – Software applications that offer recommendations to consumers based on that consumer's pattern of behavior and consumption and drawn from analyses of the consumption behavior of a large aggregate of consumers.

CompuServe – An online information service that provides access to the Internet, e-mail, instant messaging, and an integrated contact list. Founded in 1969 as a timesharing service, CompuServe is one of the oldest online services. It was the first to offer e-mail in 1979 and offered online chat a year later.

Electronic mail (e-mail) – A store and forward method of composing, sending, storing, and receiving messages over electronic communication systems.

File transfer protocol (FTP) – The protocol for exchanging files over the Internet. FTP is most commonly used to download a file from a server using the Internet or to upload a file to a server.

Genie – An online information and bulletin board service that closed its doors at the end of 1999. The General Electric Network for Information Exchange (Genie) was set up as a joint venture between General Electric and Ameritech in

1985. Its roundtable discussions, chat lines, games, and Internet access attracted a niche of science fiction aficionados as well as horror and fantasy writers.

Google AdSense – A contextual ad network that gives web publishers the opportunity to serve advertisements on their web site in return for commissions on a cost-per-click (CPC) basis.

Google AdWords – Cost-per-click (CPC) advertising. The advertiser pays only when users click on their ad. It has features that allow them to control their costs by setting a daily budget for what they are willing to spend per day. AdWords sponsored listings are also being shown on Google's partner sites.

Hyperlink – Commonly referred to as a link, this is a text string (i.e., sequence of characters) or an image in an electronic document that serves as a user-activated switch that causes another, predetermined location in the same page or document or on an entirely different page to display on the screen.

Internet – A global network connecting millions of computers. Unlike online services, which are centrally controlled, the Internet is decentralized by design.

MIDI – Musical instrument digital interface (MIDI) is a protocol designed for recording and playing back music on digital synthesizers.

MP3 – MPEG-1 Audio Layer-3 is a standard technology and format for compressing a sound sequence into a very small file (about one-twelfth the size of the original file)[4] while preserving the original level of sound quality when it is played.

Napster – A controversial application that allowed people to share music over the Internet without having to purchase their own copy on CD. After downloading Napster, a user could gain access to music recorded in the MP3 format from other users who are online at the same time.

Online service – An organization that provides access to the Internet as well as proprietary content. Before the Internet became widely used by the general public, all online services were self-contained organizations known for their unique mix of databases and resources.

P2P – On the Internet, peer to peer (referred to as P2P) is a type of transient Internet network that allows a group of computer users with the same networking program to connect with each other and directly access files from one another's hard drives. Kazaa and Limewire are examples of this kind of peer-to-peer software.

Web 2.0 – The popular term for advanced Internet technology and applications including blogs, wikis, real simple syndication (RSS), and social bookmarking. O'Reilly Media and MediaLive International originally coined the expression in 2004, following a conference dealing with next-generation web concepts and issues.

---

[4] Depending upon the selected bit rate.

REFERENCES AND FURTHER READING

Brown, Janelle. (1998). Blame it on Rio: Netheads love the mp3 digital-music format. Why does the music industry hate it so much? Salon.com, http://archive.salon.com/21st/feature/1998/10/28feature.html (October 28).

Brown, Janelle. (1999a). MP3 crackdown: As the recording industry "educates" universities about digital music piracy, students feel the heat. Salon.com, www.salon.com/tech/log/1999/11/17/riaa/index.html?source=search&aim=/tech/log (November 17).

Brown, Janelle. (1999b). MP3: Here, there, everywhere. The latest digital music players let you play MP3s on your home stereo, in your car or on the run—but are they any good? Salon.com, www.salon.com/tech/review/1999/12/14/mp3 (December 14).

Brown, Janelle. (2001). The music revolution will not be digitized. Salon.com. http://archive.salon.com/tech/feature/2001/06/01/digital_music/index.html.

Business Week. (2007). Universal Music takes on iTunes. www.businessweek.com/print/magazine/content/07_43/b4055048.htm?chan=gl.

ComputerWorld Honors. Areosmith/Geffen Award and case study. www.cwhonors.org/search/his_4a_detail.asp?id=2213.

ComputerWorld Honors. MIDI/Music Forums of CompuServe Award and case study, www.cwhonors.org/search/his_4a_detail.asp?id=2214.

Factmonster. Internet timeline, www.factmonster.com/ipka/A0193167.html.

Grossman, Lev. (2006). Time's person of the year: You. www.time.com/time/magazine/article/0,9171,1569514,00.html.

Guglielmo, Connie. (2008, April 04). Apple moves past Wal-Mart as leading music retailer in US. *The Tennessean Newspaper*, section e, p. 3.

Hormby, Thomas. (2007). History of iTunes and iPod. *The iPod Observer*. www.ipodobserver.com/story/31394 (May 10).

Information Week. A brief history of web 2.0. www.informationweek.com/1113/IDweb20_timeline.jhtml.

Kaplan, Carl. (1999). In cout's view, MP3 player is just a 'space shifter.' *New York Times* (http://partners.nytimes.com/library/tech/99/07/cyber/cyberlaw/09law.html).

Nashawaty, Chris. (1994). They byte the songs. *Entertainment Weekly*, http://www.ew.com/ew/article/0,,304033,00.html.

O'Reilly, Tim. www.oreillynet.com.

Salon.com. (1999). Rio portable MP3 player holds up in court. Music industry loses case invoking federal piracy law. http://archive.salon.com/ent/log/1999/06/16/rio/index.html.

Silverstein, Jonathan. (2006). iTunes: 1 billion served. ABC News, http://abcnews.go.com/print?id=1653881 (February 23).

Spellman, Peter. (2001). The Musicians Internet. Berklee Press Publications, Boston, MA.

Webopedia. Definitions. www.webopedia.com.

www.go2web20.net.

CHAPTER 2

State of the Market

TRENDS IN INTERNET USE

This chapter looks at the state of the market and current trends in Internet use. The discussion includes a description of the market, an overview of the state of e-commerce, and a look at specific trends in music sales online, including digital downloads.

Who's Using the Internet?

According to the Pew Internet Project's research in 2007 and 2008, 75% of U.S. adults use the Internet, with a whopping 92% of young adults (ages 18 to 29) online. Sixty-nine percent said they had used the Internet within the past 24 hours. Ninety-three percent of adults with household income above $75K use the Internet, whereas only 55% of adults making less than $30K a year are online. Internet users are also more likely to be well educated. Only 47% of adults have high-speed access at home.

Ninety-three percent of teens use the Internet, and more of them than ever are using it for social interaction. Sixty-four percent of teenagers create some kind of content for the Internet, usually through social networking sites. Girls continue to dominate most elements of content creation with some 35% of all teen girls blogging, compared with 20% of online boys, and 54% of web-using girls post photos online compared with 40% of boys who use the Internet. Boys, however, are twice as likely to post videos on the Web. Older teens, especially girls, are more likely to use social networking sites. Almost half of teens who use social networking sites visit the sites either once a day (26%) or several times a day (22%).

Among adults, e-mail and search engine use are the most popular online activities, yet nearly three-fourths of Internet users have made an online purchase. Thirty-four percent have listened to music online, whereas 27% have

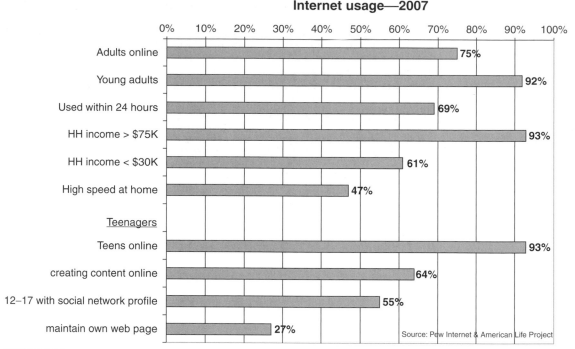

FIGURE 2.1
Internet usage among demographic groups.

downloaded music files. An equal number (27%) state they have shared files from their computer with others.

TRENDS IN E-COMMERCE

Music is a dream product for e-commerce—it is one of only a handful of products that can be sold, distributed, and delivered all via the Internet. Many products are sold online as retailers shift some marketing efforts from retail stores or mail-order catalogs to online stores. Of course, this is a double-edged sword. The ease with which music can be transferred over the Internet has also led to the ongoing crisis of illegal file sharing. Online commerce is just an extension of the mail-order business that has been around since the days of the first Sears-Roebuck catalog. In the 1960s and 1970s, credit cards and toll-free numbers helped to expand the mail-order industry. In the mid-1990s, as *online* became a household word and Internet use skyrocketed, consumers were still somewhat reluctant to purchase products online for a number of reasons: credit card security, lack of trust in the online retailers, inability to see the product before ordering, and clunky, complex storefronts and shopping carts. In 1992, CompuServe offered its users the chance to buy products online. In 1994, Netscape introduced a browser capable of using encryption technology, called secure socket layers (commonly referred to as SSL), to transmit financial information for commercial transactions online. Then as shopping interfaces improved, more established

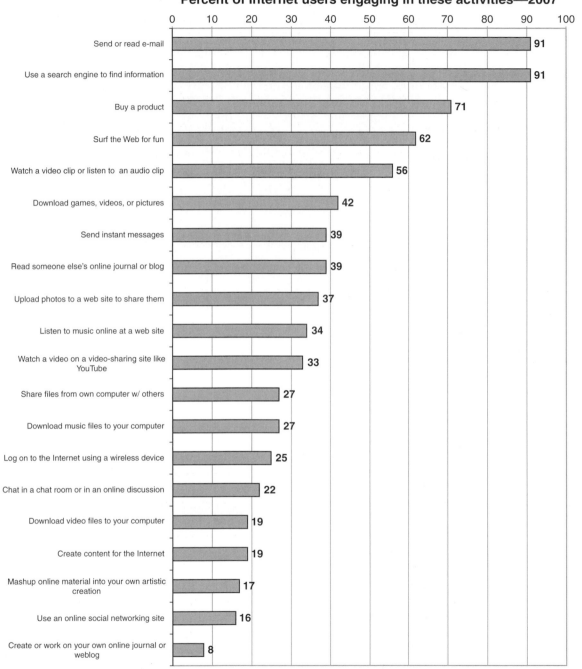

Percent of Internet users engaging in these activities—2007

Activity	Percent
Send or read e-mail	91
Use a search engine to find information	91
Buy a product	71
Surf the Web for fun	62
Watch a video clip or listen to an audio clip	56
Download games, videos, or pictures	42
Send instant messages	39
Read someone else's online journal or blog	39
Upload photos to a web site to share them	37
Listen to music online at a web site	34
Watch a video on a video-sharing site like YouTube	33
Share files from own computer w/ others	27
Download music files to your computer	27
Log on to the Internet using a wireless device	25
Chat in a chat room or in an online discussion	22
Download video files to your computer	19
Create content for the Internet	19
Mashup online material into your own artistic creation	17
Use an online social networking site	16
Create or work on your own online journal or weblog	8

FIGURE 2.2
Percentage of Internet users engaged in online activities.

stores began an online presence, alternate payment methods evolved, and consumers became more comfortable with making purchases online. In 1995, two online retail giants, Amazon and eBay, were introduced.

In January 2008, Nielsen Online reported that globally, 875 million consumers had shopped online, which is more than 85% of Web users. This was up by 40% from 2006. South Korea had the highest percentage of online shoppers, with 99% of Internet users, followed by the United Kingdom, Germany, and Japan, all with 97%. In the United States, 94% of Internet users purchase products online. The top-selling online retailers for the 2007 holiday season were eBay, Amazon, Target, Wal-Mart, Best Buy, Circuit City, Sears, ToysRUs, Overstock.com, and JC Penney. Table 2.1 shows the most popular products to purchase online.

Table 2.1	Most Popular Online Purchases—2007*
Books	41%
Clothing/accessories/shoes	36%
Videos/DVDs/games	24%
Airline tickets/reservations	24%
Consumer electronics	23%
Music	19%

*Percent of responses when consumers were asked what items they had purchased online in the past 3 months.
Source: Nielsen.

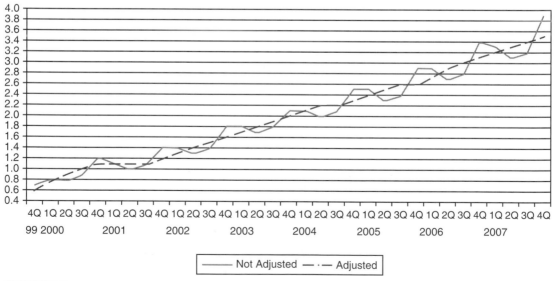

FIGURE 2.3
U.S. Census Bureau E-Commerce Retail Sales. (Source: U.S. Census Bureau.)

Sixty percent of online shoppers used a credit card for purchases in 2007, whereas 25% have used PayPal. According to Nielsen, shoppers tend to stick with shopping on sites they are familiar with, and 60% said they buy mostly from the same site.

MUSIC SALES TRENDS

In an effort to monitor recorded music sales and determine trends and patterns, the industry in general and SoundScan in particular have come up with a way to measure digital album sales and compare them with music sales in previous years. When SoundScan first started tracking digital download sales, the unit of measurement for downloads was the single track, or in cases where the customer purchased the entire album, the unit of measurement was an album. But this did not give an accurate reflection of how music sales volume had changed, because most customers who download buy songs a la carte instead of in album form. In an attempt to more accurately compare previous years with the current sales trend, SoundScan came up with a unit of measurement called *track equivalent albums* (TEA), which means that 10 track downloads are counted as a single album. Thus, the total of all the downloaded singles is divided by ten and the resulting figure is added to album downloads and physical album units to give a total picture of "album" sales. Here is an example of how this works from Billboard.biz.

> When albums are tallied using the formula of 10 digital track downloads equaling one album, the 582 million digital track downloads last year translates into 58.2 million albums, giving overall albums a total of 646.4 million units. The overall 2006 total of 646.4 million is a drop of 1.2% from 2005's overall album sales of 654.1 million.
> **Ed Christman, Billboard, January 4, 2007**

Having established the TEA as a new unit of measurement, industry trends show the following: U.S. album sales have continued to slide every year from a high in 2000 of 785 million units to 646.4 million units in 2006 (including TEA). (Without the addition of the TEA, album units in 2006 were 588 million.) This represents a drop of 1.2% from the 2005 overall album sales of 654.1 million.

By the close of 2007, album sales were down 15% from the previous year and down 20% for the holiday season. With track-equivalent albums factored in, the slide was 9.5% from 2006. The one bright spot for physical units has been online sales. Meanwhile, retail stores continued to reduce the amount of shelf space they devoted to recorded music as online sales and digital downloads continued to erode at the physical unit market share. Sales of physical albums fell nearly 19%. Digital album sales reached the 50 million mark in 2007, a 53.5% increase over 2006. Digital album sales accounted for 10% of total album sales (without TEA included).

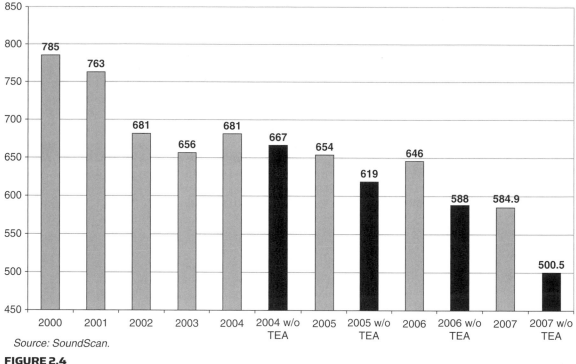

US Album Sales (in millions)

Source: SoundScan.

FIGURE 2.4
U.S. Album sales in millions.

FIGURE 2.5
2007 Global digital sales
by channel. (Source: IFPI.)

2007 Global Digital Sales by Channel

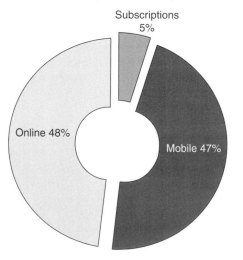

Subscriptions 5%

Online 48%

Mobile 47%

Sales of song downloads grew at 45% over 2006, but this increase was less than the 65% increase recorded a year earlier. In the United States, more than 844 million digital tracks were sold (SoundScan) compared to 582 million in 2006. When all music formats are included (music videos, singles, albums, digital, etc.), the U.S. industry fell 14.3% in 2007.

The International Federation of the Phonographic Industry's (IFPI) Digital Music Reports for 2007 and 2008 outline some trends in global online music sales, including the following highlights:

- More than 500 online music services were available in more than 40 countries.
- Portable player sales totaled around 120 million in 2006, an increase of 43% on the previous year. Portable player sales grew another 15% in 2007 to 140 million.

- Record companies' digital music sales are estimated to have nearly doubled in value in 2006 and were up another 53% in 2007 to reach 1.7 billion legally downloaded tracks.
- Social networking sites exploded in popularity, and advertising-supported models such as imeem emerged as a potentially exciting revenue stream for record companies. In 2007, all four major labels worked out agreements with imeem to provide full-length streaming of much of its catalog in exchange for a percentage of ad revenue.
- Ringtone sales seemed to peak in 2007 as users began to turn to other sources and *sideloading* for ringtones and ringtunes. In January 2008, Nielsen reported, "There were 220 million ringtone purchases in 2007 resulting in sales of $567 million, and ringtone sales spiked 22 percent in the last week of the year. 91 percent of ringtone sales were master-tones, and the top 100 mastertones sold 65.1 million, accounting for 30 percent of ringtone sales." In the United States, ringtone sales fell from 10% to 9% of overall mobile content. According to Jupiter Research, ring-tone's share of the overall mobile content market in Europe was down to 29% of mobile content in 2007, compared to 33% the previous year (Shannon, 2007).

Sales of music downloads have continued to grow, although that growth is slowing from the rapid pace earlier in the decade. In 2007, digital sales in the United States accounted for 30% of all music sold according to the IFPI (2008), but only 23% according to SoundScan. And in South Korea, digital sales accounted for over 60%. With sales revenue of $2.9 billion globally, the digital market grew 40% in 2007, up from $380 million in 2004. In market share, digital sales have grown from 2% of the market to 15% in the same four-year time period.

The number of licensed tracks available for sale online increased from 1 million in 2003 to more than 6 million in 2007. Digital sales, especially album sales,

	Table 2.2	Top 5 Digital Markets, Sales by Channel, 2007		
			Online	Mobile*
1	United States		67%	33%
2	Japan		9%	91%
3	UK		71%	29%
4	South Korea		63%	37%
5	Germany		69%	31%

*Includes full tracks and ringtones.
Source: IFPI.

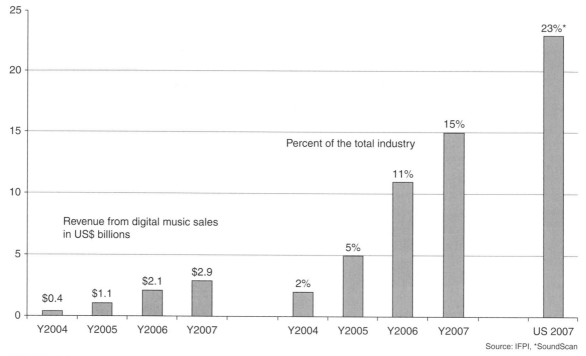

FIGURE 2.6
Global digital music sales 2004–2007. (Source: IFPI.)

have been stronger for catalog titles than their CD counterpart. In 2008, *The Wall Street Journal* commented:

> [D]igital-album sales have consistently been more weighted to catalog and "deep catalog" items (generally speaking, releases more than 18 months old) than sales of physical CDs, and catalog and deep-catalog sales have shown stronger growth.
>
> **Fry (2008)**

In 2004, catalog products accounted for 46% of all digital music sales compared with 35% for physical products. This was at a time when the replacement cycle for CDs was at an end. In the late 1980s and early 1990s, labels enjoyed record profits as consumers sought to replace their vinyl and cassette music collections with the improved digital CD format. As a result, catalog sales ran as high as 50% of units sold. Although the conversion from CD to MP3 or other portable format has generated some interest in catalog sales, consumers are not repurchasing music they already own on CD. Instead they are rounding out their catalog collection by cherry-picking songs they want in their library. The digital downloading format has, however, allowed major record labels to reissue recordings that previously had been deleted from the active catalog because, now, the cost of making them available is low.

Digital stores offer consumers a far greater virtual shelf space than the largest traditional brick and mortar stores. This means that a broader range of repertoire, including specialist, vintage or hard-to-find recordings is now available to fans.

IFPI Digital Music Report (2008)

The switch from CDs to digital downloading varies when separated out by genres. According to Nielsen SoundScan, digital album sales for 2007 were as follows: the genres of alternative, rock, soundtracks, and electronic showed an active downloading market, with fans of these genres more likely to select to download an album rather than buy the CD. Fans of country, R&B, rap, and Latin are more likely to buy CDs than to download albums.

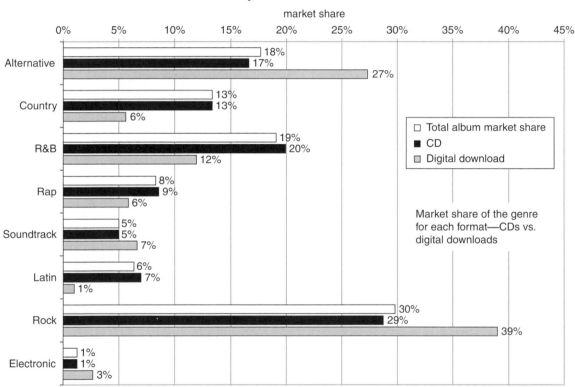

FIGURE 2.7
U.S. album sales by format for CD and digital. (Source: SoundScan.)

Business Models

A la carte downloads remain the dominant business model, with iTunes leading the way. In the United States, iTunes surpassed Amazon and Target in 2007 to become the third largest music retailer, behind Wal-Mart and Best Buy. By early

2008, iTunes had moved to first place among music retailers. Lack of interoperability between services and devices has hampered the development of the digital music market.[1] Labels have addressed this concern by offering digital rights management (DRM)-free downloads, first on a trial basis and then a larger rollout. Subscription-based services are not as popular (although these services grew 63% from mid-2005 to mid-2007), but they are also plagued by interoperability issues. Labels, artists, and services have been experimenting with offering free, advertiser-supported music, either for streaming or downloading.

Headway has been made on the streaming aspect of ad-supported music, with services such as MySpace, Bebo, imeem, YouTube, and LastFM striking licensing deals with the majors in exchange for a portion of the advertising revenue. In January 2008, CBS-owned LastFM announced it had licensing agreements with all major labels and 150,000 indies to provide full-track streaming music to fans. The basic service is advertiser supported and free to consumers, but consumers are limited to listening to a song three times before being prompted to sign up for the subscription version.

UPDATE: A correction. Though Qtrax claimed its service will be endorsed by the four biggest music labels—distributing music from all of them— three of the four big studios have denied signing any deal. So Qtrax's announcement of broad music industry support—which would have been unprecedented—doesn't seem to be true at all. Warner Music, EMI and Universal deny having licensing deals, and the fourth, Sony, isn't commenting.
Media Money with Julia Boorstin, www.cnbc.com/id/22894228, January 29, 2008

Less developed are the advertiser-supported music downloading systems, with startup QTrax stumbling out of the gate at the 2008 MIDEM show by announcing the service was in place only to have the major labels refute the claim that licensing had been established. Under this model, advertisers pick up the tab for the cost of the download. Consumers must watch an advertisement for the sponsor's product in exchange for the download. Since then QTrax has struck deals with several of the majors, making up for the false start. Meanwhile, SpiralFrog has had mixed success with its advertiser-supported downloads. The company lost $6.7 million dollars in 2006 and was still in the red in 2007. As of January 2008, Universal Music Group was the only major to have a licensing deal in place with SpiralFrog.

---

[1] Initially, labels sought to protect their music by including copy protection on music provided to legal downloading services so that the files could not be copied and shared with other consumers. As a result, different standards and platforms were developed that prevented the seamless transfer from one type of platform (such as Ipod) to another (Zune). This process is referred to as digital rights management.

On the SpiralFrog web site, music lovers can get free, legal and unlimited downloads. Anyone can access and download files from a library of more than one million songs and four thousand music videos. All you need to do to access this library is to fill out a simple and quick registration form. The company generates its revenues from relevant, targeted advertising.

Raju Shanbhag, www.TMCNet.com

TRENDS IN MUSIC DELIVERY AND MARKETING

The trend in the industry continues with music sales moving away from the CD and toward music downloading services. The attraction is in the a la carte or cherry picking that consumers prefer over purchasing an entire album. The fact that the CD has maintained the market share that it currently enjoys can be partially attributed to the belief that some consumers sometimes prefer to own music without digital rights management restricting how and where they can place copies of that music. The majority of CD music is not coupled with *copy restriction software*. The fact that more legitimate music downloads are becoming available without DRM protection may further erode CD sales in the coming years. For 2007, paid song downloads jumped to 844.2 million units in 2007, up from 581.9 million in 2006. Nine individual songs exceeded 2 million in download sales for 2007, up from just one song the previous year. Forty-one songs passed the 1 million mark compared with 22 songs the previous year.

For 2007, downloads:

- The top-selling download artist: Fergie, with 7.54 million downloads
- The most downloaded track: "Crank That" by Soulja Boy Tell 'Em, with 2.7 million tracks
- The most downloaded album: *It Won't Be Soon Before Long* by Maroon 5, with 252,000 units.
- Top-selling ringtone: "Buy U A Drank" by T-Pain, with 2.3 million units sold

In January 2008, Stuart Dredge published a list of 30 trends in digital music for 2008 in *Tech Digest*. Some of the more notable developments include the following:

1. *Advertiser-funded music.* Dredge stated that we are already comfortable with advertiser-sponsored television and radio programming, magazines, newspapers, and web sites, so it is not a great leap to extend this to music. SpiralFrog was an early entrant into the concept but has yet to make the business model successful. Dredge stated, "Nobody has found the secret formula for making ad-funded music pay off yet, but everybody is trying to." imeem has managed to strike licensing deals with all the major labels to provide full track streams in exchange for a percentage of advertising revenue. The imeem site then offers links to purchase music on Amazon and iTunes and provides widgets to post playlists on several social networking sites.

2. *Music downloads for console games.* The concept here is that once a consumer has purchased a video game, he or she can update and alter the soundtrack to include fresh new music purchased and downloaded from the Internet.

3. *Music search engines.* This point should also include music recommendation sites. The plethora of *collaborative filtering* sites for finding new music should improve the outlook for online music sales and consumption and the discovery of new music in the long tail.

4. *Choosing your own pricing.* Other artists are looking at what Radiohead initiated in 2007 when they offered their new album for download and asked fans to pay whatever they thought the music was worth. AmieStreet has a concept where the price of a track rises with its popularity.

5. *Record labels taking on iTunes.* Labels are nervous about Apple's market share of the downloading business and its unwillingness to restructure the payment model. The brainchild of UMG's Doug Morris, TotalMusic, made its debut in 2008.

6. *DRM-free music.* This trend started in 2007 when EMI announced it would start offering a DRM-free version of its catalog on iTunes for $1.29 per track. The company eventually dropped the price to the same as its other tracks. Meanwhile, Amazon launched a DRM-free MP3 store in September 2007 but has been slow to set up licensing agreements with a couple of the majors to offer their catalogs. The trend continues with more stores offering tracks that are not player-specific. This trend may actually speed the demise of the CD market. Some consumers are steadfastly continuing to buy and rip CDs so they can have the freedom to load music on to several noncompatible devices.

7. *Music identification goes interactive.* Digital radio offers listeners track information that can lead to increased sales. One of the consistent complaints from radio listeners for the past several decades has been the inability to identify and thus purchase the music they hear on the radio. Several features have been developed to remedy the situation. In addition to digital radio with the LED readout, mobile phones now offer a music ID service, which allows you to dial in a code and hold your phone up to the music source, and the service identifies the music; soon users will have the ability to click through and purchase. Another mobile application, Cliq, lets you check out the most recent playlists of radio stations and click through to buy tracks.

8. *The rise of mobile music.* Asia and Europe are already ahead of the United States in terms of mobile music, but with the introduction of the iPhone in 2007 and several other multimedia third-generation phone sets, mobile music is set to take off. Nokia has already launched and is now offering a music store. Sony Ericsson has plans to open one in 2008. *Sideloading* makes music on mobile devices more economical as consumers opt to rip and load their previously purchased music onto an iPhone or other device rather than subscribe to a mobile music service.

9. *Bands giving away their music for free.* Without ties to major labels, established bands have the option of using their music—even an entire album—as a promotional tool to be given away for free. The plan is to build a larger fan base and make up the difference in the sale of concert tickets and swag.

10. *The Net Neutrality policy is challenged. Internet service providers* (ISP) have long held to the policy of providing access to the Internet without filtering or editing the content that users access. This has allowed illegal file sharing to thrive. But as more ISPs get into the content business, their interest in protecting their own content assets calls for greater scrutiny of abuse of copyright law.

11. *Unexpected companies distributing digital music.* We have already seen Wal-Mart and Amazon dive in to the digital music market. Starbucks has now teamed up with iTunes to sell music—sometimes offering *exclusives* (CDs or special versions of albums available only in that chain's stores). In the future, there may be other retailers who are more interested in generating store traffic than turning a profit on digital downloads. It's the old *loss leader* game in digital format.

12. *Music-based widgets.* Since Facebook opened its doors to third-party developers in early 2007 and MySpace followed suit, music-based widgets have been popping up everywhere. SNOCAP enabled any musician with a MySpace account to sell digital downloads on its artist page, but it has been struggling financially as of late and was subsequently purchased by imeem. Other widgets allow artists to use their fans to virally spread their music using a variety of widgets that allow them to pass along music and videos.

13. *Social networking goes musical.* The latest generation of social networking sites are centered around music and videos, allowing users to upload, mash up, listen to, and share music and other forms of electronic entertainment.

14. *Ticket sales go mobile.* What TicketMaster did for buying tickets online, the next generation of concert ticketing is doing even more conveniently by going entirely mobile, including the use of the mobile phone as the ticket to get in to the show. "Mobi-tickets" sends a bar code to a mobile phone as a picture message, which can then be scanned by a typical point-of-sale system at the concert gates. This marketing relationship with consumers' cell phones can then be expanded by generating impulse buys on special mixes and outtakes available for a limited window of time for download to the portable communication device. It also opens the door for marketing text messages directing consumers to products. There is even discussion of tying in to a customer's physical location so that appropriate marketing messages can be delivered that are location specific. For example, consumers shopping in a particular mall may be sent text messages notifying them of a special offering or giveaway item at one store in that mall for the next 30 minutes.

CONCLUSION

The Internet continues to siphon off recorded music sales and is accounting for a larger percentage of total retail sales in many sectors. Record labels are finding new, creative ways to sell music as the role of the CD as the primary source of revenue for recorded music diminishes. New business models are being developed that generate revenue from advertising on music-related sites, advertiser-sponsored music downloads, and using music as a promotional tool to sell other products such as concert tickets or portable music devices. Record labels have instituted something called a "360-degree model," where they are responsible for and share in all revenue streams in an artist's career. Then the music can be used as a tool to generate revenue elsewhere that is shared between the artist and the label. As these models develop and as record labels scramble to remain relevant, recorded music will start generating profits in new and unusual ways.

GLOSSARY

Digital rights management (DRM) – Copyright protection software incorporated into music files so that they cannot be shared freely with peers or copied to more than the maximum number of specific portable devices.

Exclusives – CDs or special versions of albums available only in that chain's stores.

Internet service providers (ISP) – An ISP (Internet service provider) is a company that provides individuals and other companies access to the Internet and other related services.

Loss leader – Products are sold at below wholesale cost as an incentive to bring customers into the store so that they may purchase additional products with adequate markup to cover the losses of the sale product.

Music-based widgets – A portable chunk of code that an end user can install and execute within any separate HTML-based web page without requiring additional configuration. End users use them to install add-ons such as iLike and Snocap to social networking pages.

Sideloading – The act of loading content on to a mobile communication device through a port in the device rather than downloading through the wireless system.

Track equivalent albums (TEA) – Ten track downloads are counted as a single album. All the downloaded singles are divided by ten and the resulting figure is added to album downloads and physical album units to give a total picture of "album" sales.

REFERENCES AND FURTHER READING

Beta News Staff. NPD: iTunes now #3 music retailer. *Beta News*, June 22, 2007, www.betanews.com/article/NPD_iTunes_Now_3_Music_Retailer/1182537375.

Christman, Ed. (2007, January 4). Nielsen SoundScan releases year-end sales data. *Billboard*. http://www.billboard.biz/bbbiz/content_display/industry/e3iXZLO0IdrWu AOeIRwz3vtYA%3D%3D

Dredge, Stuart. (2008, January 4). 30 Trends in digital music: the full list. *Tech Digest*, http://techdigest.tv/2008/01/30_trends_in_di_6.html.

DMW Daily. (2008, January 3). U.S. album sales down 9.5%, digital sales up 45% in 2007, www.dmwmedia.com/news/2008/01/03/u.s.-album-sales-down-9.5%25,-digital-sales-45%25-2007.

E-Commerce as a revenue stream. http://newmedia.medill.northwestern.edu/courses/nmpspring01/brown/Revstream/history.htm.

Ecommerce news. (2008, January 3). Holiday online shopping up 19 percent, www.ecommercenews.org/e-commerce-news-010/0284-010308-ecommerce-news.html.

Fry, Jason. (2008, January 27). Beyond the album: 2007 brings new signs that the era of a beloved musical form is ending. *Wall Street Journal*.

IFPI Digital Music Report: 2008. www.ifpi.org.

Pearce, James. (2003). Picture this: Tickets on your cell phone. CNetNews.com, www.news.com/2100-1026_3-5103968.html.

Pew Internet and American Life Project. www.pewinternet.org.

Shanbhag, Raju. (2008). Music download site SpiralFrog announces traffic milestones. TMCnet. www.tmcnet.com/news/2008/01/28/3233924.htm (January 28)

Shannon, Victoria. (2007, December 31). Global market for cell phone ring tones is shrinking. *New York Times*. http://www.nytimes.com/2007/12/31/business/31ringtone. html?ex=1199682000&en=1d8772ac7ea7a942&ei=5058&partner=IWON

CHAPTER 3
Overall Music Marketing Strategy

Marketing an artist or any recorded music product involves a marketing plan with several elements; the resources available for marketing and the particular marketing goals of the project dictate the relative importance of each element. For instance, a local band with limited resources should probably focus more on developing a market in the geographic area where the artist performs, whereas an international star should focus more on the mass media and wide distribution. The components currently being used to market artists include the following:

- Publicity
- Advertising
- Radio promotion
- Retail promotion
- Music videos
- Grassroots marketing
- Internet marketing
- Tour support
- Special markets and products

PUBLICITY

Publicity consists of getting exposure for an artist in the mass media that is not in the form of advertising. In other words, publicists are responsible for getting news and feature coverage for an artist as well as appearances on television. These tasks include, but are not limited to, the following:

- Press releases
- News stories
- Feature stories
- Magazine covers

- Photos
- Television appearances
- Interviews

The one exception is radio airplay. The publicist is not responsible for getting songs played on the radio. That job would fall to the radio promoter (described in a later section of this chapter).

For publicity tools, publicists rely on press releases, biographies (*bios*), *tear sheets* (copies of previous articles), a *discography*, publicity photos, and publicity shots. It is the publicist who writes most of the copy for these items and prepares the *press kit*, although other branches of a record label may actually use the press kit in their marketing functions. Publicists are also responsible for media training the artist, for setting up television appearances and press interviews, sending out press releases and holding press conferences. They target a variety of media vehicles, including:

- Late night TV shows (Letterman, Leno, Conan O'Brien, *Saturday Night Live*, etc.)
- Daytime TV shows (Oprah, Regis and Kelly, etc.)
- Morning news shows (*Good Morning America*, etc.)
- Local newspapers (weeklies and dailies)
- National entertainment and music magazines
- Trade publications such as *Billboard* and *Radio and Records*
- Online e-zines and blogs
- Local TV shows (in support of concert touring)
- Cable TV shows (including music television channels, but for news and feature items, not airplay)

PR Newswire is an online service designed to disseminate press releases to the appropriate media outlets. Journalists looking for news releases on a particular topic can turn to this resource. For music publicists, it can be a convenient clearinghouse for getting a press release to the right journalists.

ADVERTISING

Advertisers determine where to place their advertising budget based on the likelihood that the advertisements will create enough of a sales increase to justify their expense. Advertisers must be familiar with their market and consumers' media consumption habits in order to reach their customers as effectively as possible.

Consumers are targeted through radio, television, billboards, direct mail, magazines, newspapers, and the Internet. *Consumer advertising* is directed toward potential buyers to create a "pull" marketing effect of buyer demand. *Trade advertising* targets decision makers within the industry, such as radio programmers, wholesalers, retailers, and other people who may be influenced by the advertisements and respond in a way that is favorable for the marketing goals. This creates a "push" marketing effect.

Advertising is crucial for marketing recorded music, just as it is for other products. The primary advertising vehicle in the recording industry for the major labels is local print sources, done in conjunction with retail stores to promote pricing of new titles, and is referred to as *co-op advertising*. But in addition to local print media, the record industry relies also on magazine, radio, television, outdoor, and Internet advertising. The impact of advertising is not easy to measure because much of its effect is cumulative or in conjunction with other promotional events such as live performances and radio airplay. *SoundScan* has improved the ability to judge the impact of advertising, but because marketing does not occur in a vacuum, the relative contribution of advertising to sales success remains somewhat of a mystery.

The most complex issue facing advertisers involves decisions of where to place advertising. The expansion of media has increased the options and complicated the decision. Table 3.1 represents a basic understanding of the advantages and disadvantages of the various media options.

Advertising is beginning to shift to the Internet for the second time. After initial efforts by advertisers to reach consumers on the Internet failed to produce the expected results, advertisers pulled out of the Internet, contributing to the

Table 3.1	Comparison of Media	
Media	**Advantages**	**Disadvantages**
Television	Reaches a wide audience but can also target audiences through use of cable channels Benefit of sight and sound Captures viewers' attention Can create an emotional response High information content	Short life span (30–60) seconds High cost Clutter of too many other ads; consumers may avoid exposure May be too broad to be effective
Magazines	High-quality ads (compared to newspapers) High information content Long life span Can target audience through specialty magazines	Long lead time Position in magazine not always certain No audio for product sampling (unless a CD is included at considerable expense)
Newspapers	Good local coverage Can place quickly (short lead) Can group ads by product class (music in entertainment section) Cost effective Effective for dissemination of information, such as pricing	Poor quality presentation Short life span Poor attention getting No product sampling

(Continued)

Table 3.1	Comparison of Media—cont'd	
Media	**Advantages**	**Disadvantages**
Radio	Is already music-oriented Can sample product Short lead, can place quickly High frequency (repetition) High-quality audio presentation Can segment geographically, demographically, and by musical tastes	Audio only, no visuals Short attention span Avoidance of ads by listeners Consumer may not remember product details
Billboards	High exposure frequency Low cost per exposure Can segment geographically	Message may be ignored Brevity of message Not targeted except geographically Environmental blight
Direct Mail	Best targeting Large information content Not competing with other advertising	High cost per contact; must maintain accurate mailing lists Associated with junk mail
Internet	Best targeting—can target based on consumer's interests Potential for audio and video sampling; graphics and photos Can be considered point-of-purchase if product is available online	Slow modem speeds limit quality and speed Effectiveness of this new media still unknown Doesn't reach entire market Internet is vast and adequate coverage is elusive

Internet bubble burst in 2001. Now, advertisers are returning to the Internet as pay-per-click advertising has offered a new way for advertisers to pay only for those ads that lead to consumer action.

RADIO PROMOTION

The impact of radio airplay on record sales and artist popularity is still the most powerful singular force for breaking new artists. The reliance on radio to introduce new music to consumers causes record marketers to focus a lot of resources on obtaining airplay. This is done through the promotion department, where radio promotion people engage in personal selling to influence radio programming. Radio program directors make the key decisions on which music is played on the radio station and which is rejected. As a result, record labels and artists lobby radio program directors to encourage them to play their music. Decisions by radio programmers are the keys to the life of a record and have become the basis for savvy, smart, and creative record promotion people to carry out this lobbying effort.

Here is how songs are typically added to the playlists for those stations that program new music:

1. A promoter who is affiliated with a record label, or an independent promoter hired by the label, calls the radio station *music director* (MD) or *program director* (PD) announcing an upcoming release. Radio music directors have specific "call times." These are designated times of the week that they will take calls from record promoters. The call times vary by station and are subject to change. For example, an MD may have call times of Tuesdays and Thursdays 2 to 4 P.M.
2. Leading up to the add date, meaning the day the label is asking that the record be added to the station's playlist, the promoter will call again to tout the positives of the recording and ask that the recording be added to the playlist.
3. The music director or program director will consider the promoter's selling points, review the trade magazines for performance of the recording in other cities, consider current research on the local audience and its preferences, look at any guidance provided by the station's corporate programmers/consultants, and then decide whether to add the song.
4. The PD will look for reaction or response to adding the song. The "buzz factor" for a song will be apparent in the call-out research and call-in requests as well as through local and national sales figures.

An important component of promoting a recording to radio is the effectiveness of the record company promotion department or the independent promoters hired to get radio airplay. This would appear to be a simple process but the competition for space on playlists is fierce. Thousands of recordings are sent to radio stations every year and the rejection rate is high because of the limited number of songs a station can program for its audience—playlists generally run under 40 songs at a time and only a few are swapped out each week. Some of the recordings are rejected from being included on playlists because they are inferior in production quality, some are inappropriate for the station's format, and many lose their label support if they fail to quickly become commercial favorites with the radio audience.

(Paul Allen in Hutchison, Macy, Allen, p. 154)

New forms of radio broadcasting, including satellite radio and Internet radio, have opened up the possibility of getting airplay for lesser-known artists. The addition of several thousand new radio stations, many with specialty formats such as unsigned bands or regional music, has increased the opportunities for emerging artists. Many of these newer formats have a less restrictive and longer playlists, so competing for airplay is not as difficult.

RETAIL PROMOTION

Radio is the most important marketing tool for influencing consumers to buy new music. So it would seem that music retailers would use radio as their primary advertising vehicle to promote their stores and product. But that is not the case. Print advertising is the music retailer's primary promotional activity. "Consumers have been trained to look in Sunday circulars for sales and featured products on most any item. Major music retailers now use this advertising source as a way to announce new releases that are to go on sale on the following Tuesday, along with other featured titles and sale product" (Amy Macy, in Hutchison, Macy, Allen, 2006, p. 213).

Promotional efforts inside record stores highlight different releases to motivate consumers to purchase the titles being promoted. It is often said that the last 10 feet before the cash register is the most influential real estate for promotional activities. Whether it is listening stations near a coffee bar outlet or posters hanging in the front window, a brief encounter within the store's walls will quickly identify the music that the store probably sells (Macy, p. 214).

Featured titles within many retail environments are often dictated from the central buying office. As mentioned earlier, labels want and often do create marketing events that feature a specific title. This is coordinated via the retailer through an advertising vehicle called *cooperative advertising*. *Co-op advertising*, as it is known, is usually the exchange of money from the label to the retailer, so that a particular release will be featured. The following are examples of co-op marketing (reprinted from Hutchison, Macy, and Allen, 2006, p. 214):

- *Pricing and positioning.* P&P is when a title is sale-priced and placed in a prominent area within the store.
- *End caps:* Usually themed, this area is designated at the end of a row and features titles of a similar genre or idea.
- *Listening stations.* Depending on the store, some releases are placed in an automatic digital feedback system where consumers can listen to almost any title within the store. Other listening stations may be less sophisticated and may be as simple as using a free-standing CD player in a designated area. But all playback devices are giving consumers a chance to "test drive" the music before they buy it.
- *POP, or point of purchase materials:* Although many stores will say that they can use POP, including posters, flats, stand-ups, etc., some retailers have advertising programs where labels can be guaranteed the use of such materials for a specific release.
- *Print advertising.* A primary advertising vehicle, a label can secure a "mini" spot in a retailer's ad (a small picture of the CD cover art), which usually comes with sale pricing and featured positioning (P&P) in store.
- *In-store event.* Event marketing is a powerful tool in selling records. Creating an event where a hot artist is in-store and signing autographs of his or her newest release guarantees sales while nurturing a strong relationship with the retailer.

As a larger percentage of record sales moves to the Internet—for physical sales as well as downloads—many well-known retailers have been forced to close their doors. Tower Records is one example. Other retailers have simply reduced the amount of floor space dedicated to recorded music and have begun to diversify if they had not already done so. In 2007, the National Association of Recording Merchandisers (NARM) awarded the large retailer of the year award to Amazon.com, the first for an online retailer.

MUSIC VIDEOS

In 1981, MTV launched the first music video channel and gave record labels a reason to produce more of the new entertainment format. Other genre-specific channels soon followed, such as Black Entertainment Television (BET), VH-1 for adult contemporary music, and Country Music Television (CMT). It became evident early on that music video exposure was beneficial to developing artists' careers and promoting their music. Stars like Madonna and Michael Jackson owe a lot of their fame to video exposure. *Telegenics* became an important aspect in artists' careers, forcing record labels to concentrate on signing artists with visual appeal.

That being said, the costs of producing a music video are astronomical, and record labels must weigh the benefits of creating one against the additional costs. Additionally, the plethora of music videos crowded the airwaves by the early 1990s, so that producing a music video did not automatically guarantee airplay on the major video outlets.

Videos remain an important promotional tool, if not for consumer promotion, then to showcase the artist to booking agents for TV shows and concerts. The rise of YouTube and other online video outlets has spurred a new kind of video format that is more edgy and an alternative to the slick Hollywood-style videos on cable TV. It's called the viral video, but virility is but one component. To begin with, many of the successful viral videos were not actually made by hobbyists but are produced to give that appearance—like the movie the *Blair Witch Project*. Then, because the videos seem to come from "the street" but have interesting qualities, they are passed around as consumers get the impression that they "found it first" and want to share this nugget with their friends. If they had the impression it was a corporate-sponsored mass media product, they would probably not bother with grassroots "sharing" because they would assume everyone would be exposed to it soon enough. A video presence is now recommended for all artists regardless of the level of their career or marketing budget.

GRASSROOTS MARKETING

Grassroots marketing, sometimes called guerrilla marketing or street marketing, consists of using nontraditional marketing tools in a bottom-up approach to develop a groundswell of interest at the consumer level that spreads through word of mouth (WOM). *Diffusion of innovations theory* discusses the role of "opinion leaders" or trendsetters who are instrumental in the diffusion of any new product, trend, or idea.

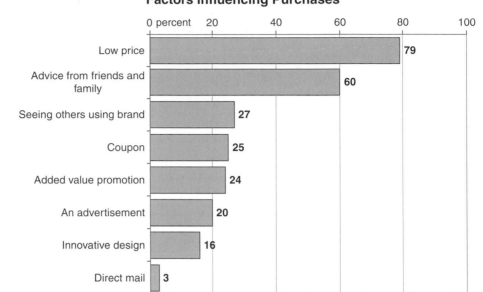

FIGURE 3.1
Factors influencing
music purchases.
(Source: Hutchison,
Macy, and Allen, 2006.)

Factors Influencing Purchases

In grassroots marketing, these opinion leaders are targeted, sometimes by hiring them to work for the artist or label, and their influence on their peers is exploited to promote new products and trends. Groups of trendsetters that become a part of the marketing establishment are called street teams. Sometimes it is enough to have the street team members adopt the new styles or consume the products visibly in the marketplace (such as drinking Sprite or wearing a new outfit). This peer-to-peer marketing is very influential, especially among younger consumers. Record labels commonly have grassroots marketing departments who manage street teams in various geographic locations around the country.

On the Internet, the concept of street team has been expanded to include using these same peer-to-peer marketing principles to reach out to consumers in chat rooms, on message boards, and in social networks (see Chapter 12).

INTERNET MARKETING

Internet marketing is the focus of much of this book. This section concentrates on the use of Internet-related sites in other aspects of the marketing plan. It is common these days for every television show, radio station, and print publication to host its own web site. Often these sites can be used for interactive marketing campaigns, such as contests. Contestants who appear on American Idol get their weekly affirmation from viewers who log on to the Internet or use their cell phone to cast their votes. Marketing for artists and record releases often incorporates a web component for each media event, whether it's to drive traffic to the media vehicle's web site or to encourage fans to sign up for information or prizes. A debut of a new video on BET can be coordinated with a promotion on www.BET.com where fans enter contests, get free music, purchase priority tickets, or participate in other promotions. In-store autograph signings by an artist can be publicized on the retailer's web site. Tour schedules can include links to each of the venues so the concert attendee can learn more about the venue and its location (where to sit, where to park, etc.).

So each aspect of a marketing plan, whether it's coverage in a particular media vehicle, tour support, or supporting an album at retail, should contain an online counterpart to make the experience interactive for consumers and to drive traffic to the media or retailer's site.

TOUR SUPPORT

Tour support is money or services that a record company provides to help promote an artist and ultimately sell more records. This is one aspect of marketing that is crucial for independent artists and those on indie labels. Generally, the marketing budget does not allow for retail product placement, radio airplay, or advertisements. So the indie artist must rely on touring to build a fan base and generate record sales. Tour support consists of contacting local media and retail in each market where the artist is to appear and providing any promotional materials and support necessary, including appearances. Aspects of tour support may be handled by publicists and the sales department of a label. The publicist is responsible for getting local media coverage and arranging for the artist to appear for interviews and impromptu performances on TV and radio. The sales department is responsible for setting up retail promotions and in-store appearances by the artist for autograph signings.

> "Touring is one of the most important ways that an independent artist can spread the word about their music within both the industry and the general public. It can serve as an excellent catalyst for radio airplay; articles, stories and reviews; retail placement, on-air performances; in-store performances and other promotional opportunities. Touring is also expensive and exhausting for most independent artists."
>
> **INDIEgo.com**

SPECIAL MARKETS AND PRODUCTS

With traditional record sales on the decline, labels and artists are always looking for new ways to make money from selling music, often referred to as special markets and special products. Getting songs included in a movie, on a TV show, or in a commercial are some examples of special markets. Compilations and samplers are special products. Retailers create "branded" CDs that are special editions available only through the sponsoring retailer. This concept is popular with all types of retailers, not just record retailers, because it draws traffic into the stores. Hallmark, William-Sonoma, and Starbucks are nontraditional retailers who use special products to draw customers to their retail establishments. In 2007, the Eagles decided to release their new album only in Wal-Mart stores. In exchange for the exclusive opportunity to offer the album, Wal-Mart spent a lot of money promoting the album that otherwise would have been spent by the record label—or, in some cases, not spent at all.

This type of marketing can now be extended to the Internet, with record companies trying out all sorts of new business models to make money from recorded music. Social networking sites such as imeem have made agreements with record labels to share advertising revenue in exchange for licensing their music to the site for streaming. Microsoft agreed to pay record labels a small token for each one of its Zune hardware players sold in exchange for offering music tracks for download that are compatible with the hardware.

CONCLUSION

The promotional aspects outlined in this chapter represent the traditional marketing methods that record labels have employed to sell their products and promote their artists. As is evidenced by the rest of this book, the old-school marketing techniques are giving way to a new type of marketing that incorporates the latest in communication and entertainment technology and that is keeping up with the rapid pace of innovation. But, old habits die hard.

GLOSSARY

Bio – Short for biography. The brief description of an artist's life or music history that appears in a press kit.

Clippings – Stories cut from newspapers or magazines.

Consumer advertising – Advertising directed toward the consumer as compared to trade advertising. Generally, this audience is reached through mass media.

Co-op advertising – An effort by two or more companies sharing in the costs and responsibilities. A common example is where a record label and a record retailer work together to run ads in a local newspaper touting the availability of new releases at the retailer's locations.

Diffusion of innovations theory – The process by which the use of an innovation is spread within a market group, over time and over various categories of adopters.

Discography – A bibliography of music recordings.

End cap – In retail merchandising, a display rack or shelf at the end of a store aisle; a prime location for stocking product.

Grassroots marketing – A marketing approach using nontraditional methods to reach target consumers.

Guerilla marketing – Using nontraditional marketing tools and ideas on a limited budget to reach a target market.

Music director (MD) – The person responsible for a radio station's playlist of songs.

Point of purchase (POP) – A marketing technique used to stimulate impulse sales in the store. POP materials are visually positioned to attract customer attention and may include displays, posters, bin cards, banners, window displays, and so forth.

Press kit – An assemblage of information that provides background information on an artist.

Press release – A formal printed announcement by a company about its activities that is written in the form of a news article and given to the media to generate or encourage publicity.

Pricing and positioning (P&P) – When a title is sale priced and placed in a prominent area within the store.

Program director (PD) – An employee of a radio station or group of stations who has authority over everything that goes out over the air.

Publicity – Getting media exposure for an artist in the mass media that is not in the form of advertising.

SoundScan – A company owned by Nielsen that is responsible for monitoring and reporting the sales numbers for recorded music. Retailers, labels, managers, agents, and other industry people subscribe to the service and retrieve the sales data online.

Street teams – Local groups of people who use networking on behalf of the artist in order to reach the artist's target market.

Tear sheets – A page of a publication featuring a particular advertisement or feature and sent to the advertiser or public relations firm for substantiation purposes.

Telegenic – Presenting a pleasing appearance on television.

Tour support – Money or services that a record company provides to offset the cost of touring and help promote the artist.

Trade advertising – Advertising aimed at decision makers in the industry, including people in radio, retail, and booking agents.

REFERENCES AND FURTHER READING

Hutchison, Tom, Macy, Amy, and Allen, Paul. (2006). Record Label Marketing. Focal Press, Oxford: UK.

CHAPTER 4

Domains and Hosts: Nuts and Bolts

WEB SITE BASICS: HOW WEB SITES ARE CONSTRUCTED

The World Wide Web uses a system of hypertext markup language (HTML) to organize text, graphics, and multimedia in an orderly fashion so that web browsers (Internet Explorer, Firefox, etc.) can make sense of the files and put together the pieces of the puzzle. On the server,[1] a web site looks like a collection of files and folders, as shown in Figure 4.1. Each web page is represented as an HTML

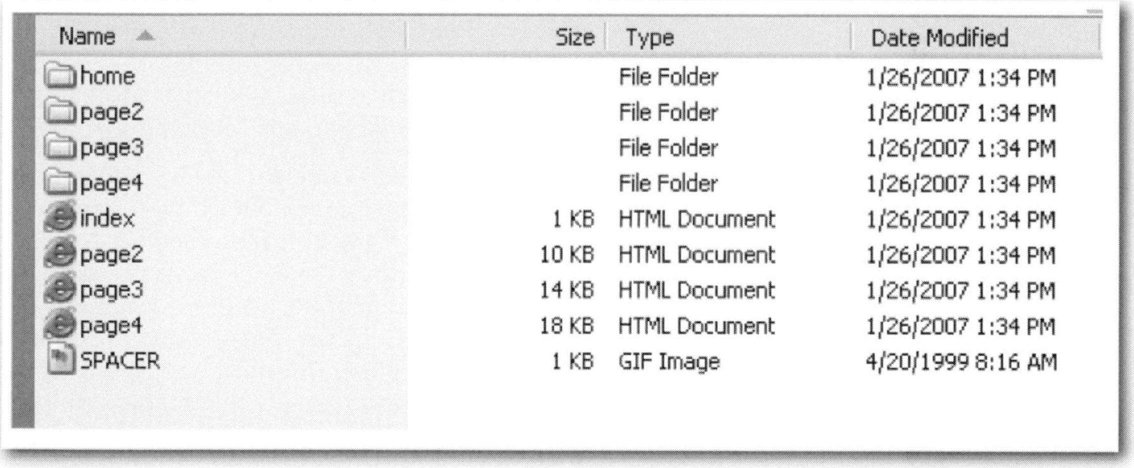

Name ▲	Size	Type	Date Modified
home		File Folder	1/26/2007 1:34 PM
page2		File Folder	1/26/2007 1:34 PM
page3		File Folder	1/26/2007 1:34 PM
page4		File Folder	1/26/2007 1:34 PM
index	1 KB	HTML Document	1/26/2007 1:34 PM
page2	10 KB	HTML Document	1/26/2007 1:34 PM
page3	14 KB	HTML Document	1/26/2007 1:34 PM
page4	18 KB	HTML Document	1/26/2007 1:34 PM
SPACER	1 KB	GIF Image	4/20/1999 8:16 AM

FIGURE 4.1
The arrangement of a web site's files and folder on the server.

[1] A Web server is the computer program (housed in a computer) that serves requested HTML pages or files.

document. The pictures and other assets for the page are located in the folder of the same name. The web browser or interface used to access and compile these pages reads HTML programming and puts the various components of the web page in the proper perspective, if all goes well.

When a web surfer types in the uniform resource locator (URL) of this location, the index page (by default) loads into the browser, picking up the page assets from the "home" folder (in this example). Links from the index page to subsequent pages connect the other HTML documents together to form the complete web site.

DOMAIN NAME

The first aspect of building a web site involves registering a domain name. The domain name is the "web address" that your visitors will become familiar with and use to access your site. The URL is the means of identifying the web site location on the Internet. An Internet address (for example, http://www.yourname.com) usually consists of the access protocol (http), the domain name (www.yourname), and the top level domain (TLD), such as .com, .org, or .net. The URL will also contain the directory path and file name, such as /myband/index.html. The general default-loading file is named *index.html* or *index.htm*. This file name should be the name of your home page, because this is the page that the web browser looks for when your customers enter in the address. The file index.html is served by default if an URL is requested that corresponds to a directory on the server where your web site resides.

Master of Your Domain

What's in a name? It is important that the URL, or Internet address, for your artist be simple and easy to remember. A long URL will confuse customers and prevent them from finding the web site. There is also more possibility of error when the customer has to enter in a long name such as http://www.cheapdomainprovider.com/~yourcustomername/personalweb/bandname/index.html.

The domain name is the single most valuable piece of real estate on the Internet. Bands that have been around since before the Internet but who were not quick to register their band name when the Internet took off are sometimes forced to use a less-than-perfect URL address. For example, the band Van Halen does not own the web address www.vanhalen.com. A fan of the band owns that site. The official band web site is www.van-halen.com, and some fans might not know or remember to include the dash. In another example, the official U.S. White House web site is www.whitehouse.gov. A whimsical company that markets political swag (slang for stuff we all get) has set up shop at www.whitehouse.org.

Many services available on the Internet will register the artist's domain name. A quick online check with these providers can determine if the artist's or band's name is available. Go for the dot-com extension if it is available, and use a domain name that is memorable and simple. This will become your online "brand."

Branding is defined as creating a distinct personality for the product (in this case, the artist, not the label) and telling the world about it. As companies rely more and more on the Internet for promotion, the dot-com name becomes a brand name. So the URL should reflect the brand that the artist wishes to promote. What if the name you want is already taken? If the artist is not yet well established, it may even be wise to rename the band or change the spelling of the artist's name to accommodate the branding that goes along with having an Internet presence. As a result of the growth of the Internet, many companies have recreated their brand to reflect the style found on URL addresses, combining words together and using capital letters to distinguish between words. Examples include names such as SunTrust bank instead of Sun Trust. This trend had been extended to include abbreviations commonly used in text messaging and similar to the shorthand found on automobile license plates, where the words *in*, *be*, *to*, and *for* are replaced by the letters and numbers N, B, 2, and 4 (e.g., NSYNC, Boyz2Men). Consider all possibilities when acquiring domain names.

It is not necessary to actually possess a physical site with the address www.band-name.com. This name can be used as the URL, and then visitors can be redirected or *forwarded* to the actual web site, which may have a longer address. In other words, the actual physical site for an artist may be http://www.recordlabelname.com/artists/the_artist/index.html but fans only need to type in www.bandname.com. It is recommended that you register derivations of the artist's name to cover visitors who may not know the correct spelling.

Consider the following tips when selecting a domain name:

1. *Make it easy to remember.* Hyphens are hard to remember and hard to communicate verbally when describing the site address to someone. If you have any peculiar symbols or spellings, be sure to emphasize that in your marketing materials.

2. *Keep it short and simple.* Also, avoid strange spellings, unless you intend to brand your artist with that spelling (such as "z" instead of "s"). Remember that words may be spelled differently outside the United States or in non-English-speaking countries. Some companies and artists like to use acronyms for the domain name (e.g., www.hsx.com for Hollywood Stock Exchange). The problem is that most of the short acronyms are already taken. Dennis Forbes (2006) reported that all of the three-letter acronyms are taken and nearly 80% of four-letter acronyms are already registered. Less than 5% of five-letter acronyms are taken.

3. *Make it descriptive of the site.* It is easier to memorize and identify a site name that relates to the content or subject matter of the site. You can add *band*, *sings*, *songs*, or some other word to the end of your artist's name, especially if the simpler domain name is taken.

4. *Use the dot-com domain, if available.* Most web surfers are accustomed to using the dot-com extension and will default to that unless you make it clear that your TLD is dot-net or dot-org. You may even want to register all variations of your domain name with the different TLDs.

5. *Use a keyword in the name, if possible.* The name of the artist is good, but adding music or band helps search engines categorize the site based on these keywords. Consider registering the name both ways.
6. Be consistent. Even if you register several variations of your domain name to cover all bases, stick with just one for your marketing materials to avoid confusing your customers. It is common to register all variations (www.arstistname.com, www.artist-name.com, www.artistnameband.com) and then forward all of these to the same domain.
7. *Your domain name should be your site name.* This is important because when people think of your web site, they will think of it by name. So have the domain name match the name of the site that appears in the masthead.

The bottom line is that the domain name is important. In the article "The Importance of a Quality Domain Name," John Stone (2006) stated, "Your domain name is your Internet phone number. Many of the same rules apply. Get a number that is easy to remember or spells your business name."

Registering Your Name

Once you have selected a domain name or determined the domains that you would like to register and use, there are several *domain name registrar* services that will register your domain name for $10 or less. The cost of registration has gone down in recent years, making it affordable to register more than one name or variation. Getting your domain name involves registering with Internet Corporation for Assigned Names and Numbers (ICANN).

Registration is for between one and nine years. Some domain registration services such as GoDaddy.com will automatically renew your registration annually and charge the cost to your credit card or PayPal account. Some web hosting services will include the cost of domain registration in the fee for hosting services. Some of the more popular domain registration services include Yahoo!, GoDaddy.com, Dotster.com, Network Solutions, and Register.com. You can also search for available domain names at these sites.

> ICANN is responsible for the global coordination of the Internet's system of unique identifiers. These include domain names (like .org, .museum and country codes like .UK), as well as the addresses used in a variety of Internet protocols. Computers use these identifiers to reach each other over the Internet.
>
> **www.icann.org**

Forwarding and Masking

Domain forwarding or *URL redirection* redirects all web traffic for a domain name that you have registered to the specific URL where the site is located. When someone types your domain name into a browser, the system will automatically forward or redirect that person to whatever host you specify that contains the folder with the web site in it. For example, writer Lynne Hayworth's web site is located at www.hutchtom.com/hayworth/index.html. Web surfers looking for that site can type in www.lynnehayworth.com and automatically be redirected to the proper site.

Usually the service provider with which you register your domain name will offer domain forwarding and *masking*. When setting up domain forwarding, it might be beneficial to employ *domain masking*, a process that keeps your registered domain name in the URL box instead of displaying the actual URL address for the server and folder where the site is located. This could eliminate some confusion for the web visitor and perhaps protect the true identity or location of the web site. With domain masking, you are requesting that the service in charge of the forwarding "mask" the actual address by keeping the registered domain name in the address bar. In addition to masking, it will be necessary to submit a masked title, keywords, and a description.

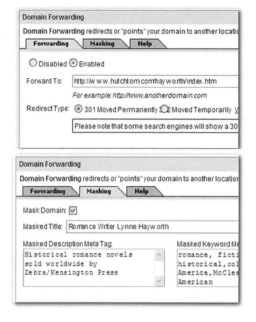

FIGURE 4.2
Example of domain forwarding.

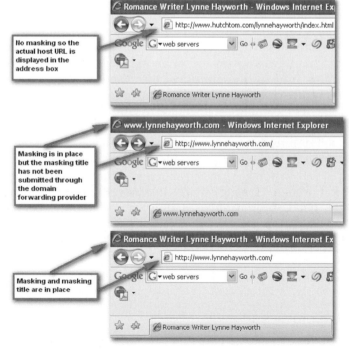

FIGURE 4.3
Example of domain masking and title masking.

WEB HOSTING

Web sites are stored on a physical server, or host. *Web hosts* are companies that provide space on a server they own for use by their clients. They also provide Internet connectivity to access that server. Most services provide basic functions for storing web pages, storing files, and offering e-mail. If a record label has its own server, it can use URL redirection to forward the artist's URL to the label's server. If the label does not have a host provider, many commercial companies offer "space" to host a web site. There are companies that offer free web site hosting. These services should be avoided for a professional site such as that for an artist or record label because the free services tack on banner advertisements and popup ads. These advertisements annoy customers and discourage them from visiting the site. It is worth the money to pay for a cleaner site, without all the ads. There are two considerations when selecting a web host: *storage space* and *bandwidth*.[2]

How Much Space?

It is important to ensure the hosting site has enough data storage space for all the functions a music-based site needs. Personal web sites, with photos, blogs, and text, use less than 5 megabytes (MB) of server space. Thirty MB of space may not be enough to hold graphics, photos, and sound files—one MP3 file can take up to 4 MB. Also, it is necessary to have enough space for expansion. An artist may only need to feature three audio files from the current album when the web site is first set up, but when subsequent albums are released, the demands on the site will also increase. Up to 5 MB are needed for graphics and web documents. A thorough site should include a bio, photos, band news, a tour itinerary, a place to sign up for a mailing list, press and media coverage (reprints), a discography, lyrics, audio files, contests and giveaways, merchandise, links to other favorite sites, video, and contact information. It is vital to include some way for fans to purchase the artist's recordings—either directly from the web site or by directing them to an online retailer. The site will require a virtual shopping cart if the artists want to sell directly from their site (discussed later in the book) and some scripting. All this can add up to quite a bit of storage space. It is recommended that at least 50 MB of storage space be allocated, with the option to trade up to a better plan if needed. Hosting service GoDaddy.com offers 5 gigabytes (GB) of space and 250 GB of bandwidth (monthly) in its economy plan for less than $5 per month.

---

[2] This is only true if a straight, static HTML site is created, which most today are not. If not, there are also other considerations to make such as installed server-side software packages. These must correspond to how the web site is created and what coding tools are used. For example, if any database tools are used (for mailing list signup, etc.), the most popular database for web processing is MySQL. MySQL must be installed on the host in order for the web site to function properly. There is also the LAMP (Linux-Apache-MySQL-PHP) versus .NET (Microsoft's standard) differences to consider. A server with the proper capabilities for each must be chosen depending upon programmer preference.

FIGURE 4.4
Servers.

Bandwidth and Uptime

Sufficient *bandwidth* needs to be allocated to accommodate numerous visitors to the site. Bandwidth is the amount of data that can be transmitted in a fixed amount of time—it is the size of the "highway." If a lot of users access a server simultaneously, sufficient bandwidth is required to transmit the information from the server to the users in a timely fashion and to avoid Internet congestion. It is wise to purchase more storage space and bandwidth than you think you will need. Some host companies are like cell phone companies: If you exceed your limit, the additional charges can add up in a hurry. The free services sometimes impose a bandwidth limit and will temporarily shut down sites that exceed this limit. That would be disastrous for a business that depends on Internet traffic for income.

Another factor is *uptime*, the percentage of time that the host is accessible to potential customers attempting to visit your web site. A hosting provider should have close to 100% uptime so that your customers can access at any time without receiving error messages.

Choosing a Host

Often a web hosting package will include other features such as e-mail accounts with the address of info@yourbandname.com, web traffic statistics, subdomain access, and database and forms management. Before deciding on a host and a hosting package, it is a good idea to determine what services you will need as the web site grows in popularity. Christopher Heng (2004), in his article "How to Choose a Web Host," listed the following criteria as important:

1. *Advertising.* Most free hosts impose advertising to offset their costs. It is not advisable to go with a site that is advertising competing or other products. The distractions are not attractive to the visitor.

2. *Amount of web space.* The popular services such as GoDaddy.com offer reasonably priced packages with 5 GB of space for less than $5 per month.

3. *FTP access, or file transfer protocol.* This is the ability to easily move files between the server and your hard drive. It is important that the software used to create the web site has the ability to "talk" to the server and engage in uploading and managing the site. If you use Microsoft FrontPage to create the web site, it is important that the host provide FrontPage server extensions. According to Kevin Spenser (1998), "The FrontPage Server Extensions are a group of programs which run on a server and allow for interaction with FrontPage Explorer when you do file/folder maintenance functions, such as moving, deleting, and renaming files and/or folders in your FrontPage webs. For example, if you move a file from one folder to another, the Server Extensions will edit all hyperlinks to and from that file to point to the correct (new) locations."

4. *Data transfer, including traffic and bandwidth.* The amount of bandwidth you are allowed to use to both upload materials and provide access for web site visitors is usually restricted by the web hosting package. Heng stated that a new site is likely to use less than 3 GB of bandwidth per month. Go Daddy's economy hosting plan allows for 250 GB of monthly traffic.

5. *Technical support.* This will allow your web site to provide reliable, consistent accessibility to visitors and prevents downtime caused by technical difficulties.

6. *SSL, MySQL, and the shopping cart.* If you plan to conduct transactions through your site, SSL (secure socket layer) guarantees encryption for credit card numbers and other sensitive information; it will be discussed in the section on e-commerce. MySQL is an open source relational database management system that allows information from online forms to be processed through a database. According to Webopedia, "A shopping cart is a piece of software that acts as an online store's catalog and ordering process." Services like PayPal offer this feature, so it may not be necessary for an artist to provide it on the artist's web site.

7. *Control panel.* This allows the webmaster to easily manage aspects of the web account features. The control panel for the Go Daddy economy hosting service is featured in Figure 4.5.

FIGURE 4.5
Economy hosting features offered by GoDaddy.com. Copyright © GoDaddy.com, Inc. All rights reserved.

CONCLUSION

The cost of setting up and maintaining a web presence is within reach for every musician, songwriter, and singer. The most important web possession is the domain name, and the domain name should be registered early in the process. There are several popular web hosting and domain registration services. Make sure the hosting plan is sufficient to serve the artist as the career develops.

The next chapter discusses web design.

GLOSSARY

Bandwidth – The amount of information or data that can be sent over a network connection in a given period of time. Bandwidth is usually stated in bits per second (bps), kilobits per second (kbps), or megabits per second (mps).

Domain forwarding – Redirecting requests on the Internet to a different Internet address. For example, domain forwarding allows multiple domain names to be registered, all of which point to the same Web site.

Domain masking – Also referred to as "masking" or "cloaking," works with web forwarding to keep your custom web address (e.g., www.yourdomain.com) in the browser address bar while visitors browse different pages in your site.

This can be used to hide the real addresses of your web pages, either because those addresses are long and complicated (e.g., http://members.bigcompany.com/~username/page.html) or because it gives your web site a more professional appearance. Your visitors will see a cleaner, more memorable address in the browser address bar.

Domain name – A name that identifies one or more Internet protocol (IP) addresses.

Control panel – Included in web hosting packages is an online web-based application that allows you to easily manage different aspects of your account. Most control panels will let you upload files, add e-mail accounts, change contact information, set up shopping carts or databases, view usage statistics, and so on.

FTP – File transfer protocol. This is the language used for file transfer from computer to computer across the Web.

Internet Corporation for Assigned Names and Numbers (ICANN) – ICANN is responsible for the global coordination of the Internet's system of unique identifiers.

Masthead – An alternate name for the nameplate of a magazine or newsletter.

MySQL – Pronounced "my S-Q-L" or my-sequel, MySQL is an open source relational database management system (RDBMS) that uses Structured Query Language (SQL), the most popular language for adding, accessing, and processing data in a database. MySQL is the most popular database system for web sites.

SSL – Short for secure sockets layer, a protocol developed by Netscape for encrypting communications on the Internet.

Top level domain – The last part of an Internet domain name—that is, the letters that follow the final dot of any domain name.

URL – Uniform resource locator, the global address of documents and other resources on the Web. The first part of the address is called a protocol identifier, and it indicates what protocol to use; the second part is called a resource name, and it specifies the IP address or the domain name where the resource is located.

Web host – A business that provides server space, web services, and file maintenance for individuals or companies that do not have their own servers.

REFERENCES AND FURTHER READING

Forbes, Dennis. (2006). Interesting facts about domain names, www.yafla.com/dennisforbes/Interesting-Facts-About-Domain-Names/Interesting-Facts-About-Domain-Names.html.

Heng, Christopher. (2004). How to choose a web host, www.thesitewizard.com.

Heng, Christopher. (2004). Tips on choosing a good domain name, www.thesitewizard.com.

Spenser, Kevin. (1998). Understanding FrontPage server extensions, hwww.abiglime.com/webmaster/articles/frontpage/042198.htm.

Stone, John. (2006). The importance of a quality domain name, http://tools.devshed.com/c/a/Domain-Name-Tips/The-Importance-of-a-Quality-Domain-Name.

www.Webopedia.com.

CHAPTER 5
Creating the Web Site

WEB SITE GOALS

Before one can begin designing a web site, it is necessary to determine the purpose of the site and outline the goals for the web site. For a musician or singer, these goals should include the following:

- Creating and reinforcing brand awareness
- Advertising and promoting products (recordings and concert tickets)
- Generating direct sales (mail order and e-commerce)
- Creating a sense of community among an artist's fans
- Creating repeat traffic

The design will need to incorporate aspects that support and enhance these goals. It should be attractive and suit the artist's image. It must also be designed to appeal to the target market.

Branding

Branding is described as creating a distinct personality for the product (in this case, the artist, not the label) and telling the world about it. Artists who create a well-known brand can parlay that into endorsement deals, an acting career, or the position as a spokesperson for a worthy cause. Artists Madonna and Missy Elliott appeared in TV ads for The Gap, Queen Latifah has moved into films, and Bono of the rock group U2 has become the spokesperson for solving the problems of Third World debt and global trade. Therefore, artist branding should be an important part of any web site. Creating an easily identifiable logo and maintaining consistency in style and design can help support the brand. This extends to the style of the web site, which should convey the image that you want to project for the artist.

Promoting and Selling Products

The web site should create a demand for the artist's products. The primary products are recorded music and tickets to live shows. Secondary products include posters, bumper stickers, and other souvenirs. Enticing the web visitor to purchase the recordings, or even just creating a desire to purchase, should be an important aspect of the site. This goal can be achieved through offering music samples on the site and information about the music and the recordings. Links to online retailers will encourage the visitor to follow through on purchase intentions, creating the impulse buy. Tour schedules, concert photos, samples of live recordings, and links to sites that sell tickets can create a demand for concert tickets. Maps to the venues also encourage concert attendance. The web site is a great place for the merchandising of T-shirts and other mementos. Photos of these items are important (see the section on e-commerce).

Sense of Community

Artist web sites are a good place for like-minded people to connect with each other. These fans probably have much in common—especially their interest in the artist. A sense of community can be created through message boards, online chats with or without the artist, and by including photos of fans at concerts. These devoted fans can become opinion leaders (the virtual street team[1]) by influencing their friends to become fans of the artist and encouraging their friends to visit the web site. MySpace has provided an excellent opportunity for fans to engage in community social behavior centered on a common interest in an artist.

The web site should also provide materials for professional journalists looking for information on the artist, as well as contact information for both journalists and booking agents. Some sites provide print quality publicity photographs and in-depth biographical information suitable for reprint in mainstream publications.

Creating Repeat Visitation

It is important to provide elements on the web site that will keep visitors coming back as new developments occur in the artist's career. Certain attributes can be included in a web site that encourage repeat visits; these are outlined in Chapter 7 These features include blogs, updated tour itineraries, contests, news, new music samples, and message boards. One entrepreneur even offers a service that provides updated horoscopes that you can post to your web site to encourage repeat traffic (www.tomorrowsedge.net).

---

[1] A street team in the music business is a local group of people, generally members of the target market, who use networking on behalf of the artist in order to reach their target market. A virtual street team serves this same purpose on the Internet and is not necessarily local.

WEB SITE DESIGN

Web site design is defined as the creation and arrangements of web pages that make up the site. It's part art and part commerce. The first page is the home page, although some web sites will have a splash page with a welcome message or graphic image that sets the tone for the web site (see the section on splash pages). Aspects of the design include content, usability, appearance, and visibility (the ability of users to find your site on the Web). A good web site is one that is attractive, uncluttered, quick to load, and easy to navigate. The site must offer something of value to the consumer—information, products, and freebies. It also helps to make the web site fun, refresh content, and give people a reason to return. Effective web sites avoid large flash programs, flashing text, animation, and large graphic files. Half of Internet users do not yet have cable modems, and a web site that is slow to load is sure to fail.[2]

The home page identifies your company or brand and extols the benefits of the site and its products. The site description, news, and a logo or image should be featured on the home page. This page should be updated frequently to reflect changes and notify the visitor of new and interesting developments. The home page often posts announcements, although a "What's New" page can also contain news briefs and updates. The home page should also contain links and navigational tools for the rest of the site. A good web page should balance text with graphics or images and sport a layout that is inviting and interesting.

Basic Design Rules

Once you have established the goals and determined the audience for your web site, it is time to decide on layout and content for the web site in general and then for each page. This is called *storyboarding*; it is when "the organized content is used to develop a diagram or map" (www.amacord.com/services/storybrd. html). The storyboarding process consists of developing a sketch of the site's structure, a detailed structural outline (which is the URL of each page listed in an organized format), and a detailed sketch of each page.

It is important to first get a sense for the overall site, how many pages will be needed, and what materials to include on each page. Content should be categorized according to user needs and organized in a way that takes into consideration the audience characteristics, their information preferences, specifications of the majority of computers, and the audience's web experience.

It might be helpful to used blank index cards, one for each web page, and list the information that should be included on each page. Then for layout, use blank letter-sized sheets to create a diagram for each page. By determining the scope of the site first, the designer can then get an idea of the navigation needs and how to set up menus and style sheets (see the WYSIWYG section).

---

[2] Some web sites offer the visitor "high-speed" and "low-speed" versions or "Flash" and "Non-Flash." The visitor can make the decision at the splash screen and proceed accordingly. Of course, this adds to the complexity and the cost of the web site.

Consider these simple rules of web design:

1. *Design web sites, not pages.* Determine the overall look of the site and the scope of information before attempting to design any one particular page. This will help the designer determine how many pages are needed, what should be on each page, and the overall consistent look of the site.

2. *Keep page size to a reasonable length.* The general rule is to design the page so that it is no longer than twice the length of the computer screen, especially for the home page. Readers do not like to scroll down and tend to become lost or discouraged by long web pages. The exception to this may be blogging, which tends to be more linear than most other web elements. When long pages are used, be sure to have plenty of bookmarks that can direct the user quickly back to the top of the page. Never use horizontal scrolling.

3. *Use appropriate graphics.* Don't load a lot of photos on one page—it will slow the loading time. Too many photos overload both the page and the eye. Make sure all photos are appropriate for that particular web page. If you want to include an album of photos, relegate it to a special photos page. Use thumbnails to present examples of photographs that can be enlarged at the click of a mouse. Web visitors are now accustomed to thumbnails but should be reminded that they can click for an enlarged version of the photo. Sometimes this is done with a "mouse-over" command, where a text message "click to enlarge" appears as the visitor runs the mouse over the thumbnail.

4. *Clearly specify the purpose of the web site and why the user should visit regularly.* The home page should have a clear, simple headline description of the site. If that is not suitable, then the masthead and dominant graphic or photo should clearly illustrate the site's purpose or products. Stick to the subject in the web site, keeping your goals in mind. If you want to include something that is not germane to the topic, consider linking to a separate web site. Keeping the site current and changing out content will encourage repeat traffic.

5. *Keep it simple and easy to read.* Nothing screams amateur more than a site that is cluttered and hard to read. Avoid backgrounds that conflict with the text and confuse the eye. Set up a color scheme that is appealing, consistent, and does not induce eye fatigue. Yellow or red text on a white or busy background never works.

6. *Balance the design elements.* Don't go too heavy on either text or graphics. Just like for a magazine layout, white space can be very appropriate when used properly. Basic principles of desktop publishing also apply to web design.

7. *Content is king.* Content should drive the design. Make your design appropriate for the theme and brand of the artist or product. Use content to give visitors a reason to explore the site and to return.

8. *Make navigation easy.* It is important to link pages in a consistent, well-planned manner that is intuitive to users. Navigation bars are usually found across the top, under the masthead, or on the left side of the page. Web visitors are used to this design, which was originally created because web pages load from the top and left side, and navigation information was sure to appear on the page regardless of the end user's browser or computer screen. Keep your navigation bar in the same location on every page, with a clearly identified link back to the home page.

9. *Keep your site up to date.* Visitors will not return to a site that looks abandoned, with old news, outdated information, out-of-style design elements, and cobwebs. If you plan to update on a regular basis, state that intention on the site and then *be sure to follow through!* If you state you will provide a weekly podcast, then be sure to have a new one available at the same time each week.

10. *Be consistent.* Keep the basic look and theme the same throughout the site. Use the same color scheme, background, and fonts on each page. Do not use many different fonts. For a more consistent look, use the same masthead on every page—with slight variations if necessary to identify individual pages. Stick with standard layouts. The reason the three-column layout is so common is that it works (Kymin).

Table 5.1	Red Flags of an Unprofessional Web Appearance
Poor browser compatibility	
Animated bullets	
Too many graphic and line dividers	
Multiple banners and buttons	
Poor use of frames	
Poor use of tables	
Too much advertising	
Large welcome banners	
No meta tags	
Under construction signs	
Scrolling text in the status bar	
Large scrolling text across the page	
Poor use of mouse-over effects	

Source: www.phoebemoon.com.

FIGURE 5.1
Example of an artist web site from www.sauceboss.com.

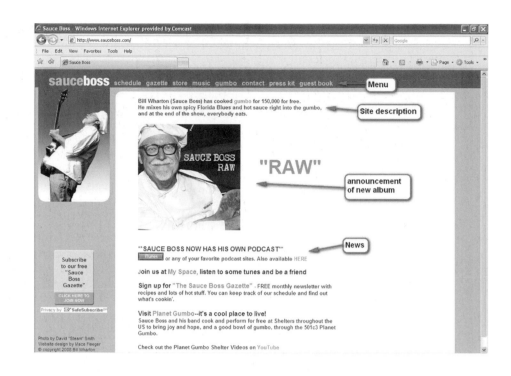

Splash Page

A *splash page* can be used to enhance the image of the web site. This is the first page a web surfer is directed to when accessing the site. Usually a splash page will have the logo and elaborate graphical layout, but not much information. The visitor is encouraged to enter the site by clicking on the page. Automatic splash pages direct the visitor to the home page after a few moments.

Do not overload the visitor with large flash programs that take a long time to load. An effective splash page has only a graphic image and the brand logo. Many web designers discourage the use of a splash page because it is just one more step or click required before the visitor gets to the true substance of the web site. However, Web Design Services India states that splash pages can be useful for entertainment web sites such as those for movies, video games, children, music, photographers, and travel sites.

If you must use a splash page, Client Help Desk (www.clienthelpdesk.com, 2004) advises one to "consider dropping a cookie into each site visitor's computer that automatically skips the splash screen on subsequent visits. Even people with the patience to deal with a splash screen once will be tested upon seeing it repeated each time they return to the site." Always offer the visitor the option to "skip the intro" and move on quickly to the home page.[3]

---

[3] Some music sites with splash screens also play music while the splash screen loads and after. While the use of music is discouraged on the splash screen or home page feature, you should also include a mute button in addition to the "skip intro" feature if you insist on including music with the splash screen (or any screen).

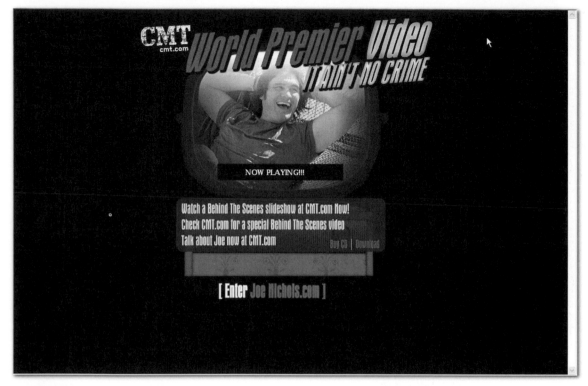

FIGURE 5.2
Splash page from Joe Nichols web site. (Courtesy of Joe Nichols and CountryWired.com.)

WYSIWYG or WYSINOT?

Many web design software programs feature a process called WYSIWYG, which means what you see (on the screen) is what you get. One of the first idiosyncrasies web designers encounter is the fact that a web site's look will vary from one computer to another. Not all web visitors use the same browser software and hardware, so what may look right on one computer may look completely different on another. Some of the variables that affect consistency are type of software browser, monitor size, monitor resolution, and the end user's font settings. Web designers must not make the mistake of assuming that a layout will look as good on other people's computer as it does on theirs. The standards for HTML agreed upon by the World Wide Web Consortium (W3C) are helping to create some consistency among various browsers, but not all users will have the latest browser version.

> You can't expect your designs to look absolutely identical on every computer. Web design is just not like that. There are situations where they can break completely—elements shift to a different place on the page or disappear completely.
>
> **Joe Gillespie, www.wpdfd.com**

The screen captures shown in Figures 5.3, 5.4, and 5.5 illustrate the variations in a web site's look based on browser and screen resolution. They all show the exact same web page.

The web page shown in Figure 5.3 was created on a *liquid layout,* where the size of the web page adjusts itself to fit the screen. On larger screens with higher resolution, the fixed elements (photos) remain the same size, whereas the liquid aspects (text arrangement) vary from computer to computer and are set up to conform to a certain percentage of the computer screen. The same web page shown on a screen with a higher resolution (Figure 5.4) has the menu bar on the right side at a distance from the main photograph. To minimize this problem, use fixed-width tables then set the elements inside specified-width boxes in the table. For years, the use of tables was popular until the introduction of cascading style sheets and universal standards for web pages.

On the high-resolution screen shot in Figure 5.4, note how the menu bar is far to the right of the photograph, rather than blended in, as was the case with the lower resolution Microsoft Internet Explorer screen shot. On the Netscape screen grab (Figure 5.5), the menu is beneath the photo rather than on the right side of the screen.

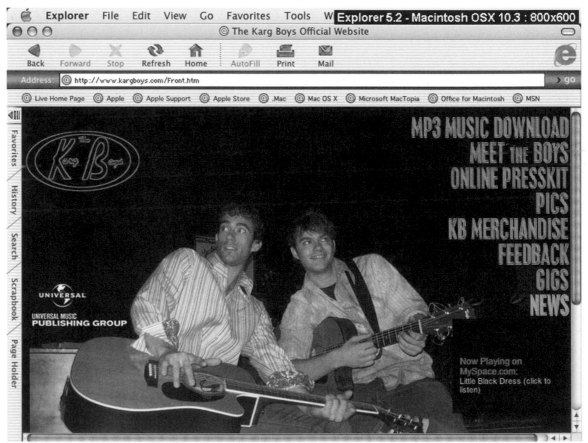

FIGURE 5.3
Example of web page in Explorer 5.2. (Courtesy of www.kargboys.com.)

FIGURE 5.4
Example of web page on high-resolution screen. (Courtesy of www.kargboys.com.)

FIGURE 5.5
Example in Netscape 4.78 on a high-resolution screen. (Courtesy of www.kargboys.com.)

It is best to check out a new web design on different systems before launching the site. There are several online services (such as www.browsercam.com) that will provide screen shots of your web site loaded with different browsers, operating systems, and monitor resolutions.

Frames versus Tables versus Cascading Style Sheets

Beginning with Netscape 2.0, web designers started using frames. The Web Style Guide describes frames as "meta-documents that call and display multiple HTML documents in a single browser window." Frames are commonly used to centralize navigation through the web site. A menu is displayed in one frame with content pages displayed in another. As a user moves through the web site, web page content changes but the navigation remains unchanged. If the navigation window has enough components, it will be displayed with a scroll bar. Many browsers cannot print frames appropriately and print each frame on a separate sheet of paper. Frames are no longer considered a recommended form of web design.

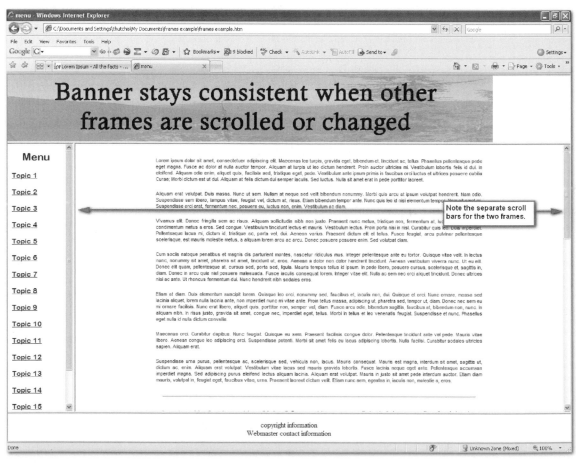

FIGURE 5.6
Example of frames page layout.

As web browsers evolved, frames then gave way to tables and nestled tables, which allowed for more content organization. Tables allow for greater control over page layout and create more visually interesting pages. Tables are used to define and separate elements in the document, such as navigation bars, masthead, side bars, text, and photos. The example in Figure 5.7 has four columns and five rows showing. Note how some of the cells have been merged to create a masthead or banner that spans more than one column or text that spans more than one row.

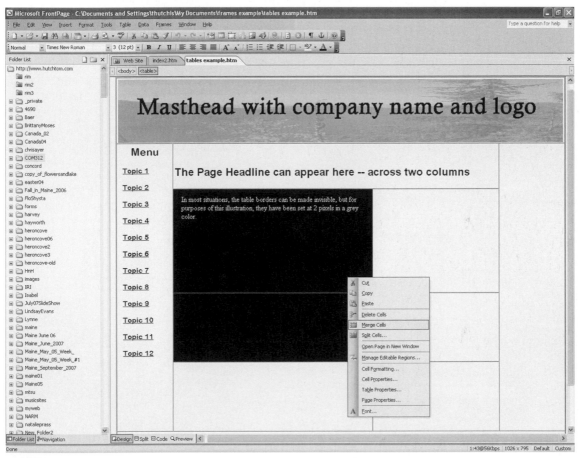

FIGURE 5.7
Example of tables layout.

In Figure 5.8, the borders of the cells have been made visible to demonstrate a tables document. By reducing the border width to zero, the cell borders become invisible while still constricting the cell content to the specified area. Cells can be created to float (expand to fill the screen) or set to a specific width and height. Cells can also have different backgrounds to set them apart from the others.

Recently, cascading style sheets have become popular. "Style sheets were developed as a means for creating a consistent approach to providing style

This example uses a four-cell table in which the two rows on the left side have been merged for a text box to extend the length of the two rows

This smaller box uses a different color scheme to set it aside from the box on the left.

The bottom box has no background color. The cells have no visible borders.

FIGURE 5.8
Example of tables with no cell borders.

information for web documents" (Wikipedia). Cascading style sheets (CSS) allow web developers to control the style and layout of multiple web pages at once, just by editing one master template (the CSS document). CSS is used to define colors, fonts, and layout, and is designed to separate document content (written in HTML) from document presentation (written in CSS). CSS has various levels, each building on the last by adding new features. Style sheets are saved in external .css files. The appearance and layout of multiple pages can be changed all at once by editing the style sheet.

FIGURE 5.9
Example of web page using cascading style sheets. (Courtesy Jo Dee Messina and CountryWired.com.)

Graphics

Graphics are an important part of any web site. There are some general rules to follow when using graphics on a web page. The three most popular formats for using graphics on a web page are JPEG, GIF, and PNG. JPEG is short for Joint Photographic Experts Group. It is good for photographs and supports 16.7 million colors. The compression actually throws out data to create a smaller file. Sharp edges may appear blurred, so JPEG is not recommended for graphics that contain sharp lines or drastic color changes. It also does not support transparency, so if you want the background of your image to be transparent, JPEG is not the format to use. A progressive JPEG file presents a low-quality image at the first moment of download, and then over several passes it improves the quality.

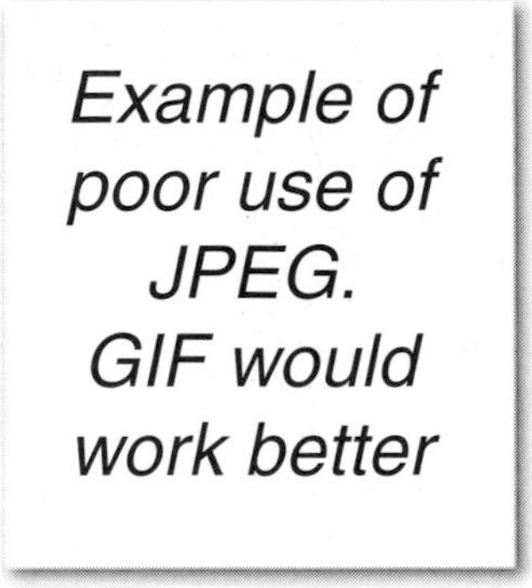

The GIF (graphics interchange format) format is excellent for graphics that have large areas of the same color. The format supports a maximum of 256 colors and thus is not the best to use for photographs. Gradual changes in color may show up as progressive rings of color changes, as evident in Figure 5.11. The GIF format supports transparency, allowing the graphic designer to remove the background of the image. The newest format, PNG (Portable Network Graphics), has images that always look great and offer good compression ratios. PNG was designed to offer the best features of the GIF format, but with millions of colors. PGN also allows for transparency, but it is not supported by older browsers.

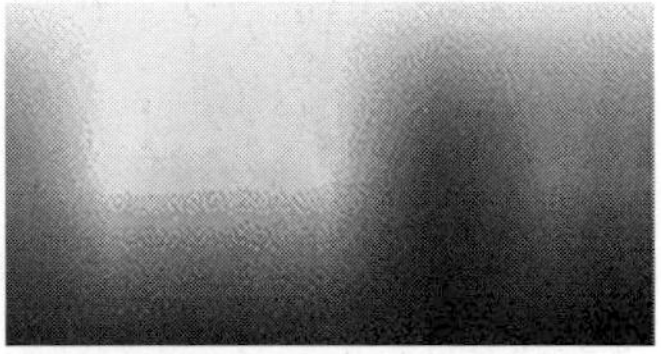

FIGURE 5.10
Example of a poor use of JPEG.

FIGURE 5.11
Example of a poor use of GIF.

EDITING GRAPHICS

Adobe Photoshop is the industry standard for image manipulation (editing graphics and photographs). Even the novice user is easily able to crop pictures, create layers, overlay text, rotate images, modify colors and contrast, convert formats, and brush out unwanted elements of the photograph. For the amateur photograph editor, there are some shareware and free programs available for basic photo editing.

In 2008, Adobe Systems launched a free online version of Photoshop called Photoshop Express. This web-based version works with any type of computer and operating system. Adobe also offers a boxed consumer version of Photoshop Express (Fehd, 2008).

The online site Picnik.com (www.picnik.com) offers users the opportunity to manipulate images online and includes widgets for photo editing your existing photos on other sites such as flickr, Facebook, Photobucket, Picasa, and Webshots. Picnik offers a free version or a premium version for a small annual fee.

FastStone offers a freeware version of its image viewer, which contains quite a few image editing and manipulation functions, such as those basic functions listed earlier, drop shadow, sepia tone, red-eye reduction, special effects, sharpen or blur, and resample compressed photos. The downloadable software is available at www.faststone.org. A commercial version is also available for a reasonable fee, and FastStone offers several other useful products.

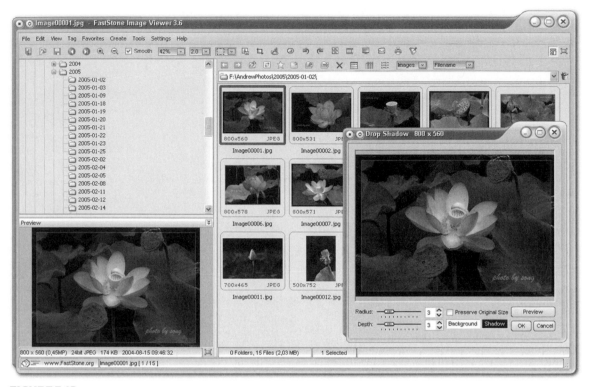

FIGURE 5.12
FastStone image editor. (Courtesy of FastStone, www.faststone.org.)

Table 5.2	Photo Editing Software
Software and URL	**Description**
Adobe Photoshop Express online www.adobe.com/products/ photoshopexpress/ ?promoid=CBTVM	The free version of Adobe Photoshop Express is available online, but not for download, and has hobbyist features but lacks the high-end professional features of the full program.
FastStone Image Viewer and Editor www.faststone.org	FastStone supports all major graphic formats including BMP, JPEG, JPEG 2000, GIF, PNG, PCX, TIFF, WMF, ICO, and TGA. It has a nice array of features such as image viewing, management, comparison, red-eye removal, e-mailing, resizing, cropping, color adjustments, and musical slide show.
GIMP for Windows www.gimp.org	GIMP is an image manipulation program, a freely distributed piece of software for photo retouching, image composition, and image authoring. It works on many operating systems.

Table 5.2	Photo Editing Software—cont'd
Software and URL	**Description**
Serif PhotoPlus www.freeserifsoftware.com/ software/PhotoPlus/default. asp	PhotoPlus is a photo editing software that enables users to fix and enhance digital photos, create stunning bitmap graphics, and even produce web animations.
Paint.net www.getpaint.net	Paint.NET is free image and photo editing software for computers that run Windows. It features an intuitive and innovative user interface with support for layers, unlimited undo, special effects, and a wide variety of useful and powerful tools.
Pixia http://park18.wakwak.com/ ~pixia/download.htm	Pixia is free, Windows-based software that allows for free painting and retouching. It features custom brush tips, drawing tools, and image color adjustments.
ImageForge www.cursorarts.com/ ca_imffw.html	ImageForge provides a set of tools for painting and editing images, photos, or other graphics. Create and edit images; acquire pictures from a scanner, digital camera; apply special effect filters; and produce photo albums and simple slide shows.
Ultimate Paint www.ultimatepaint.com	Ultimate Paint is a full-featured 32-bit Windows graphics program for image creation, viewing, and manipulation.
XnView www.download.com/ XnView/3000-2192_4- 10067391.html?tag=lst-0-5	XnView supports red eye correction, crops and transforms JPEG images losslessly, generates HTML pages and contact sheets, and provides batch conversion and batch renaming.
Saint Paint Studio www.download.com/Saint- Paint-Studio/3000-2192_4- 10066321.html?tag=fd_sptlt	The Saint Paint Studio paint package is designed to be the essential base tool for editing photos, web graphics, icons, images, and animations.
For more photo editor software, try www.WM4MB.com	www.photo-freeware.net www.download.com www.softpedia.com

In an article titled "Top 8 Free Photo Editors for Windows," Sue Chastain mentioned the following free photo editors. Her list has been expanded to include a couple other programs that are either shareware or offer a free trial period.

CREATING THUMBNAILS

Thumbnails of photographs are popular on web pages. A thumbnail is a small version of a picture that opens a larger graphic file of that picture when the user clicks on it. Most web design programs will automatically create thumbnails with a simple command, but the concept is simple enough to create with a bit of HTML text. First, open the photograph in any photo editing program. Reduce the size of the photograph to the ideal thumbnail size (maybe 150 pixels wide by 200 pixels high). Be sure to resample the resized photograph (thumbnail) for best results. Save the new photograph under a different name so that it won't overwrite the original. A good rule of thumb is to use the same name but add "_small" to the name or "_thm" so you can distinguish it from the original. Then when creating the web page layout, include the smaller thumbnail-sized photo on the page. Create a hyperlink that will lead to and open the larger graphic file. The HTML may look something like that shown in the accompanying box.

HTML LANGUAGE FOR THUMBNAILS

```
<body>
<p><a href=http://www.kargboys.com/images/KBCover.jpg target="_blank">
<img src="http://www.kargboys.com/images/KBCover_small.jpg" alt="Karg Boys cover"
width="150" height="160"></a></p> </body>
```

The alt="Karg Boys cover" denotes text that will be used to fill in if the image does not load into a user's browser. The target command will open the photograph in a new window. If you want it to open in the same browser window, the target command would be target="_self".

There are also several shareware programs that will create thumbnails quickly if you have many photos to process, including Easy Thumbnails. Easy Thumbnails is a popular free utility for creating accurate thumbnail images and scaled-down/ up copies from a wide range of popular picture formats (www.fookes.com/ezthumbs).

FIGURE 5.13
Example of photo and thumbnail.

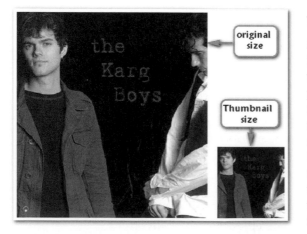

WEB DESIGN SOFTWARE

Most software programs provide WYSIWYG editing, meaning you can manipulate components on the web page while seeing exactly what the finished product will look like (theoretically). This has made web designing much easier, especially for the novice web developer. In addition to WYSIWYG, most programs also provide a code-based view for HTML tweaking. For most popular web development programs, it is recommended that the novice user start with ready-made templates (see the section on templates), many of which are available for sale or for free by third-party developers.

The most commonly used programs to create web sites are Adobe's Dreamweaver and Microsoft's Front Page, although Front Page has been replaced by Office SharePoint Designer 2007 and Expression Web, which are partially based on FrontPage technologies.

ConsumerSearch.com states that "**Dreamweaver** is the best (albeit expensive) choice of web authoring tools for Windows and Mac users. Adobe Dreamweaver CS3 is the darling of the professional web design community for its powerhouse features, including strong support for Cascading Style Sheets (CSS) and all major scripting languages" (www.consumersearch.com/www/software/web-design-software).

FrontPage from Microsoft has been another commonly used professional web development program, for both the novice and professional, and it allows for FTP uploading of web pages and files. Microsoft has phased out FrontPage but continues to support versions that are still in use in the marketplace.

Microsoft's newer web development software is **Expression Web**, which is the design tool that is part of the Microsoft Expression suite, along with Expression Graphic (for graphic design) and Expression Interactive (for application design). The newer suite of programs works more seamlessly with other standards such as XHTML (extensible hypertext mark-up language—a hybrid of XML and HTML) and CSS coding for style sheets. In fact, the program is designed to encourage the use of CSS standards. Bloggers and reviewers compare it more closely to Dreamweaver than to the earlier FrontPage program and say it was built from the ground up rather than being an upgrade of FrontPage.

CoffeeCup is an inexpensive, easy-to-use program for beginners, with detailed tutorials. The software is available with many add-on components to create extra features. The HTML editor allows for WYSIWYG editing in the "visual editor" mode or HTML editing in the "code editor" mode. In the visual editing mode, you can drag and drop images, text, and tables, and edit on the fly.

Web Studio 4.0 is the priciest of the midrange software. It is easy to use for the newbie and includes add-ons like a library of templates, photos, buttons, and special effects. Its features include a WYSIWYG design with drag-and-drop components, FTP uploading, and a built-in graphics generating tool, but it lacks an HTML code view for editing, and adding external HTML code can be a challenge.

Web Plus, according to ConsumerSearch, has features comparable to WebStudio. It offers a range of templates, smart objects, and automatic navigation bars and has a free version, Web Plus SE. The company offers templates, wizards, and tutorials to help newcomers create attractive web sites. Web Plus X2 is the commercial version.

Web Easy Professional 6.0, according to ConsumerSearch.com, "features WYSIWYG with drag-and-drop, FTP uploading, CSS support, real simple syndication (RSS) feed support, and multimedia support including Flash animation and scripting. There is a library of templates and thousands of clip-art images, as well as a built-in graphics generating tool. However, it lacks a spell checker and has no HTML editing capability—Web Easy Professional is strictly WYSIWYG." Web Easy is highly rated by Top Ten Reviews.com. Web Easy Express 6.0 is a freeware version.

WYSIWYG **Web Builder** is a low-cost, easy-to-use, but limited design program. It features drag-and-drop components including flash and forms. Web Builder features online tutorials, and a fully functional 30-day trial version is available.

PC Magazine describes **SJ Namo WebEditor 6 Suite** as "an affordable entry-level program that can grow along with a user's skill level and development needs." The magazine gives it a high rating and touts its easy-to-use tools. David Nevue, in his book *How to Promote Your Music Successfully on the Internet*, stated that it is his personal favorite of the WYSIWYG editors and he personally uses it. There is a free trial version, and the full version sells for less than $100. The negatives listed by both *PC Magazine* and CNET review are that it lacks sufficient tutorials and the features are buried inside dialog boxes.

SiteSpinner is a software program mentioned in Nevue's book as recommended by one of his readers. Top Ten Reviews.com rates it as average, with nice text and image editors, all for less than $50. SiteSpinner offers tutorials, user guides, and online support, as well as a 15-day free trial period.

Nvu is an open-source free program with the basic tools needed to create a web site and an interface that is reminiscent of earlier FrontPage programs. It claims to support many of the same easy-to-use features that make Dreamweaver and FrontPage so popular. Nvu includes some more advanced bells and whistles like JavaScript coding and CSS support. It supports forms and has a file management system (FTP). It provides a WYSIWYG edit view, a code view, and preview mode, similar to FrontPage, and it is compatible with both Windows and Macintosh platforms.

WEB DEVELOPMENT SOFTWARE

The major software packages are available at many online and physical retail locations.

Microsoft Expression Web, www.microsoft.com/expression/products/overview.aspx?key=web

Front Page, http://office.microsoft.com/en-us/frontpage/default.aspx

Adobe Dreamweaver, www.adobe.com/products/dreamweaver

CoffeeCup, www.coffeecup.com/software

Nvu, http://nvudev.com/index.php

Namo WebEditor, www.namo.com

WebStudio, www.webstudio.com

SiteSpinner, www.virtualmechanics.com

Web Easy, www.v-com.com/product/Web_Easy_Pro_Home.html

Web Builder, www.wysiwygwebbuilder.com

WebPlus, www.freeserifsoftware.com/software/WebPlus/default1.asp

TEMPLATES ANYONE?

Templates are predesigned web pages that feature basic elements (placeholders) such as text boxes, backgrounds, buttons, a header, and images. The idea is that you take the template and replace the generic elements with your own. For the novice web designer, using a ready-made template may be a good place to start. Even if you end up throwing out much of the original template and replacing it with your own features, it creates a baseline to work from. The downside of using a template is that your web site might not look unique if others also use the template with little modification. But generally, the template can serve as a guideline because designers with more experience created them. It also provides some of the basic elements such as buttons, backgrounds, dividers, and text boxes. In her article "Why and How to Use Templates Effectively," Jacci Howard Bear suggested that templates can save time, provide consistency from page to page, and be a less expensive alternative to hiring a designer. She suggested the following steps to modify a template and make it your own:

1. Select the right template, one that closely suites the subject matter so that fewer modifications need to be made.
2. Change the graphics from the generic stock photos and graphics to your own.
3. Change fonts to your own, but keep them suited for the style and image of the site.
4. Change text formatting, perhaps adding bullet points or subheadings.
5. Change the template color scheme. The background, text colors, and other elements can be changed to reflect the image of the site.
6. Change the template layout. Although the original purpose of having a template is to provide a ready made web page design, alterations such as flipping to a mirror image or swapping out sections can help to personalize the template.

Lorem ipsum dolor, what? Many templates come with dummy or placeholder text of nonsense designed as filler until the web designer can replace it with actual text for the web site. In her article "Lorem ipsum dolor," Jacci Howard Bear suggested working with the placeholder text to experiment with font types and colors and background colors to get an idea what the final product will look like and to give you an idea how much text to write for the web page. You can cut and paste the placeholder text over and over until it approximates the length of the replacement text. It comes in handy if you need to design a web page or a newsletter before the actual article copy is ready for insertion. Just don't forget to replace the dummy text before publication. You can generate your own dummy text at www.webdevtips.co.uk/webdevtips/codegen/clickgo.shtml.

Many of the web site packages listed in this chapter contain some basic templates. For those with no prior web design experience, it is good practice to experiment with those templates before deciding on a look and layout for your web site. Of course, these packaged templates are the least unique because everyone using that software has access to them. Many third-party companies offer templates. Some are software specific and allow for a more seamless interaction with the program. Others are general HTML coded templates that are designed to work reasonably well with many web design software programs. Some templates are available for free, whereas others charge a fee for either one template or a series. A good place to start looking is TemplatesReview (www.templatesreview.com).

Template Monster is a site that is often cited in blogs and reviews (www.templatemonster.com). The service gets high marks from reviewers and users. You can use the search function to specify what types of templates you are looking for (music) and what feel you want to them. For less than $50, the search returns several possibilities. The site also offers a host of free sample templates and free clip art. **4Templates** is another site that offers low-cost templates, starting at less than $22 (www.4templates.com). **FreeWebsiteTemplates** offers a wide selection of templates at no cost, including Flash templates (www.freewebsitetemplates.com).

ELEMENTS AND CONTENT FOR AN ARTIST WEB SITE

An artist's web site should contain elements that help achieve the goals set out at the beginning of this chapter: branding, promoting products, creating a sense of community and generating repeat traffic. As a part of the branding process, the overall look of the site should reflect the taste of the artist and the expectations of the target market.

An artist's web site should contain the following basic elements:

1. *A description and biography of the artist.* The home page should have some information or description, but a separate biography page should be created for in-depth information about the artist.
2. *Photos.* Promotional photos, concert photos, and other pictures of interest. This can include shots of the artist that capture everyday life, photos of the fans at concerts, and other photos that reflect the artist's hobbies or interests.

3. *News of the artist.* Press releases, news of upcoming tour dates, record releases, and milestones such as awards. This page should be updated often, and outdated materials should be moved to an archive section.

4. *Information about recordings.* Discography and liner notes from albums to increase interest in the recorded music of the artist. Comments from the artist on the recordings. Information on the recording process.

5. *Song information.* Lyrics and perhaps chord charts (again to increase interest in the recorded music).

6. *Audio files.* These may be located on the purchase page to encourage impulse purchases. They may be 30- to 45-second samples or streaming audio rather than downloadable files, to protect against piracy.

7. *Membership or fan club signup page.* Allows visitors to sign up for your newsletter or to access more exclusive areas of the site. This will help you build an e-mail list and allow for more control over content posted on message boards in restricted areas of the site.

8. *Tour information.* Tour dates, set lists, driving directions to venues, photographs/video from live performances, touring equipment list.

9. *E-store.* Merchandise page for selling records, T-shirts, and other swag.[4]

10. *Contests or giveaways.* To increase repeat traffic and motivate fans to visit the site. These can be announced at concerts.

11. *Links.* To other favorite sites, including links to purchase products or concert tickets, venue information, the artist's personal favorites, e-zines (online magazines), the artist's MySpace page, and other music sites. Ensure that all your offsite links open in a new window so the visitor can easily return to your site.

12. *Contact information.* For booking agencies, club managers, and the media. This could also include print-quality images for the press.

13. *Message board or chat rooms.* This allows the fans to communicate with one another to create a sense of community. This can be an area restricted to members only.

14. *Blogs.* Recently blogs have become popular on the Internet. A blog is simply a journal, usually in chronological order, of an event or a person's experiences. Maintaining a blog of the touring experience is one way to keep fans coming back to the web site to read the most recent updates to the journal. It also gives fans a sense of intimacy with the artist. Video blogging (or vlogging) is also becoming popular.

15. *Printable brochures or press kits.* Electronic versions of any printed materials that the artist uses in press kits and to send out should be made available on the web site. (See Appendix 1.) On the link to retrieve these items, be sure to mention if they are in the PDF format—some older browsers and slower modems lock up when attempting to open a PDF file in a browser window. Sonicbids is an online site that offers users the ability to create professional-looking electronic press kits for artists (www.sonicbid.com).

---

[4] The e-store can be either integrated into the web site or feature a link to the artist's products on a third-party selling site.

By including these elements you will cover the goals of the web site, generate traffic and repeat visits, and provide an around-the-clock source of information and entertainment for fans. Coupled with an aggressive web promotion campaign, a well-designed web site can increase the visibility and popularity of an artist at any stage of the artist's development and career. Once you have determined what content should be included on a web site, decisions must be made regarding the design and layout—what colors, fonts, images, and so on should be used.

The best advice is to scour the Internet to find web sites and web components that are appealing and serve as examples and inspiration for building the perfect site. If a design expert is to be employed, these examples will illustrate to the designer what is expected of the new site. For the do-it-yourselfers, templates are available for most web design programs. Some are offered as part of the software package, some are offered as free downloads, and some are offered for sale (often for a small fee) through commercial software web sites. It may be worth spending a few dollars on a template with a professional and contemporary look. All templates can be modified or customized—the idea is to keep the elements that work and replace those that do not. A good web site can also be the product of evolution. Each time the site is updated, it is tweaked with minor improvements until it finally has the intended look of professionalism and success.

In conclusion, the web site is considered the home base from which promotions are launched. All marketing and promotional materials and campaigns can then direct fans to the web site for more of what they like about the artist. But the web site is just the beginning of the Internet presence for an artist. Chapter 7 addresses how to maximize the web site to increase traffic and sales. Chapters 10 through 12 outline how to promote the web site and the artist on the Internet.

Creating Content for the Web Site

An artist web site is one part press kit, one part electronic store front, and one part social networking site. One of the most challenging aspects of creating a new web site is building up the assets that will be used on the site. The earlier section on basic design rules discussed how to take the content and design a site using storyboarding. The previous section included a list of potential items that may be incorporated into the web site, but how are these assets developed? The publicist is the best person to create these assets. He or she knows how to craft stories and select photographs that depict the artist in the image that best suits the artist's career. Do not leave it to the webmaster to create and select these assets—this is not the webmaster's area of expertise.

If the artist has a press kit, that is a good place to start—using electronic versions of text files and graphics. These items should be developed in advance of designing the web site and should be developed for both online and offline use. An artist biography is usually written by a trained journalist or publicist, who interviews the artist, reads up on the artist's background, understands the market, and then creates a compelling story via the bio. A professional photographer should be hired to take publicity photos. Publicity shots are not the same as a publicity photo. A publicity shot is one taken backstage with other celebrities or at events. The publicity photo is the official photographic representation of the artist. Publicity photos should be periodically updated to keep current with styles and image. But once a photo is released to the public, it is fair game for making a reappearance at any time in the artist's career, even for artists who have moved on and revamped their image.

The press release is another asset that should be incorporated into the web site. Press releases are a standard tool in public relations, one that works better than letters or phone calls (Spellman, 2000). The press release is used to publicize news and events and is a pared-down news story. Here are some examples of when a press release should be used:

1. To announce the release of an album
2. To announce a concert or tour
3. To publicize an event involving the artist or label
4. To announce the nomination or winning of an award or contest
5. To publicize other newsworthy items that would be appealing to the media

The press release should be written with the important information at the beginning. Today's busy journalists don't have time to dig through a press release to determine what it is about. They want to scan the document quickly to determine whether the information it contains is something that will appeal to their target audience.

The press release needs to have a slug line (headline) that is short, attention grabbing, and precise. The purpose or topic should be presented in the slug line. The release should be dated with contact information including phone and fax numbers, address, and e-mail. The body of text should be double spaced. The lead paragraph should answer the five W's and the H (who, what, where, when, why, and how). Begin with the most important information; no unnecessary information should be included in the lead paragraph (Knab, 2003). In the body, information should be written in the inverse pyramid form—in descending order of importance.

CONCLUSION

A web site should be created after determining the goals of the site, the image of the brand, and the expectations of the visitor. The elements for inclusion should be decided in advance, and the site should be planned before work begins on development. Assets or content for the site should be developed independently with consideration given for elements that will be used both on the web site and in other promotional materials. Publicists should work with web designers and marketers to ensure that the web site meets the goals. In addition to assets, web designers[5] and developers need tools, including a web design program and an image editing program.

[5] There is a difference between web designers and web developers. A designer would not need a web design program, they work only with the graphic design and image editing tools, mostly Photoshop. The developer would be using the site design tools such as Dreamweaver, although most pro developers use other things. Occasionally these two people are one and the same, but in most pro web shops they're at two different desks.

APPENDIX 1

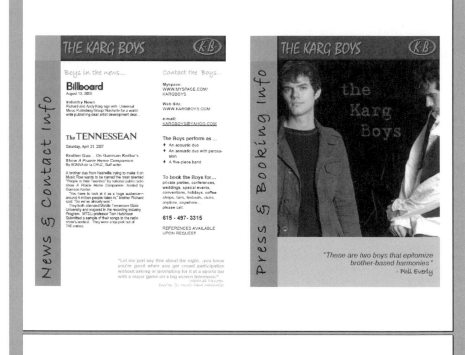

FIGURE 5.14
Printable press kit available from the artist's web site as PDF file. (Courtesy Karg Boys.)

GLOSSARY

Banner – A typically rectangular advertisement placed on a web site, either above, below, or on the sides of the web site's main content and linked to the advertiser's own web site.

Blog – Short for web log, a blog is a web page that serves as a publicly accessible personal journal for an individual. Typically updated daily, blogs often reflect the personality of the author.

Branding – Creating a distinct personality for a product (in this case the artist, not the label) and telling the world about it.

Cookie – A message that a web server gives to a web browser. The browser stores the message in a text file. The main purpose of cookies is to identify users and possibly prepare customized web pages for them.

CSS – Short for cascading style sheets, a new feature being added to HTML that gives both web site developers and users more control over how pages are displayed. With CSS, designers and users can create style sheets that define how different elements, such as headers and links, appear. These style sheets can then be applied to any web page.

FTP – File transfer protocol, the protocol for exchanging files over the Internet.

GIF – Pronounced *giff* or *jiff*, GIF stands for graphics interchange format. GIF is limited to 256 colors and is more effective for scanned images such as illustrations than it is for color photos.

Home page – The main page of a web site. Typically, the home page serves as an index or table of contents to other documents stored at the site.

JPEG – Short for Joint Photographic Experts Group, and pronounced JAY-peg. JPEG is a compression technique for color images, used for most photographs on the Web.

Liquid layout – Layouts that are based on percentages of the current browser window's size. They flex with the size of the window.

Navigation bar – A set of buttons or graphic images typically in a row or column used as a central point that links you to major topic sections on a web site.

PDF – Portable document format (PDF) is a file format that has captured all the elements of a printed document as an electronic image that you can view, navigate, print, or forward to someone else. PDF files are created using Adobe Acrobat, Acrobat Capture, or similar products (whatis.com)

PNG – Short for Portable Network Graphics, and pronounced ping, a new bit-mapped graphics format similar to GIF.

Splash page – The page on a web site that the user sees first before being given the option to continue to the main content of the site. Splash pages are used to promote a company, service, or product or to inform the user of what kind of software or browser is necessary to view the rest of the site's pages.

Storyboarding – A storyboard is an expression of everything that will be contained in a program or web site—what menu screens will look like, what pictures (still and moving[6]) will be seen including when and for how long, and what audio and text will accompany the images, either synchronously or hyperlinked.

Thumbnail – A miniature display of a page or photo that can be enlarged when clicked. This enables a web page design that can hold many small images. The thumbnail is simply a smaller version of the picture connected by a hyperlink to the larger photo.

Virtual street team – A group of Internet marketers, generally selected from the target market, who actively promote the artist on the Web, generally from the perspective of the artist's avid fans.

Webmaster – A person or group of people responsible for the design, implementation, management, and maintenance of a web site.

WYSIWYG – Pronounced *WIZ-zee-wig* and short for "what you see is what you get." A WYSIWYG application is one that enables you to see on the display screen exactly what will appear when the document is printed or published to the Web.

REFERENCES AND FURTHER READING

Bear, Jacci Howard. Lorem ipsum dolor, http://desktoppub.about.com/cs/pagelayout/a/lorem.htm.

Bear, Jacci Howard. Why and how to use templates effectively, http://desktoppub.about.com/cs/pagelayout/a/use_templates.htm.

Chastain, Sue. Top 8 free photo editors for Windows, http://graphicssoft.about.com/od/pixelbasedwin/tp/freephotoedw.htm.

CNET Web Editor Reviews. Namo WebEditor Suite 2006, http://reviews.cnet.com/web-graphics/namo-webeditor-suite-2006/4505-3637_7-31752255.html.

ConsumerSearch.com. Web design software reviews, www.consumersearch.com/www/software/web-design-software.

Fehd, Amanda. (2008, March 27). Adobe launches free version of Photoshop. *The Tennessean*, business section, p. 1.

Gillespie, Joe. www.wpdfd.com.

Knab, Christopher. (November 2003). How to write a music-related press release, http://www.musicbizacademy.com/knab/articles/pressrelease.htm.

Kymin, Jennifer. Basics of web design, http://webdesign.About.com/od/webdesigntutorials/a/aa070504.htm.

Moon, Phoebe. www.phoebemoon.com.

---

[6] Technically, these are called static and dynamic. In web page terms, "static" is a page with fixed text and standard images; "dynamic" is a page with variable information, moving or rotating pictures (or animation), and possibly some visitor interaction.

Nevue, David. (2005). How to promote your music successfully on the Internet, www.promoteyourmusic.com.

PC Magazine Software Reviews. SJ Namo WebEditor 6 Suite, www.pcmag.com/article2/0,1759,1607976,00.asp.

Spellman P., (2000, March 24). Media power: Creating a music publicity plan that works: Part 1, www.harmony-central.com./Bands/Articles/Self-Promoting_Musician/chapter-14-1.html.

Top Ten Reviews. SiteSpinner, http://website-creation-software-review.toptenreviews.com/sitespinner-review.html.

Top Ten Reviews. Web Easy, http://website-creation-software-review.toptenreviews.com/web-easy-professional-review.html.

Web Design Services India. www.webdesignservicesindia.com.

Web Graphics. www.sph.sc.edu/comd/rorden/graphics.html.

Web Style Guide. http://webstyleguide.com/page/frames.html.

www.pcmag.com/encyclopedia_term.

www.amacord.com/services/storybrd.html.

www.webopedia.com.

www.theclienthelpdesk.com.

CHAPTER 6

HTML and Scripts

HTML: THE BASICS AND WHY YOU NEED THEM

Hypertext markup language (HTML) is the predominant authoring language for the creation of web pages. HTML defines the structure and layout of a web document by using a variety of tags and attributes to denote formatting of certain text as headings, paragraphs, and lists. HTML is written in the form of tags bracketed by the greater than and less than symbols, such as < tag >. Most tags come in pairs, the opening tag is listed as <tag>, and the closing tag is </tag>, which denotes the end of the previous command. For example, if you wanted to italicize a word in a sentence, you would precede the word with the tag for italics <i> and follow the word with the end tag </i>. Failure to include the end tag would result in everything from that point forward being presented in italics. A web visitor's browser examines the HTML for instructions on how to display the graphics, text, and other multimedia components. Tutorials can be found at www.w3schools.com.

> When someone types in a URL or clicks on a Web page link, the browser requests a document from a Web server via the hypertext transport protocol, or HTTP. The server then sends the document back to the user, which is displayed on the browser. The things that are contained in the document (text, photos, audio and video files, etc.) were all put there using HTML structure.
>
> **www.answers.com**

All HTML documents start with the command <html>. This lets the browser know to read and interpret the commands as HTML. The last tag on the web page should be the end tag </html>. This tag tells your browser that this is the end of the HTML document. Beneath the <html> tag at the top of the page, you will find header information between the <head> and </head> tags. This information is not displayed on the page. The head element contains information about the document, some of which helps search engines catalog and describe the site. Head tags include those shown in Table 6.1.

Here is an example of the source code for a heading for a web page head that was shown in Chapter 5 (Figure 5.1):

```
<!doctype html public "-//w3c//dtd html 4.0 transitional//en">
<html><head>
<meta http-equiv="Content-Type" content="text/html; charset=iso-8859-1">
<meta name="DESCRIPTION" content="Bill Wharton (Sauce Boss) has cooked gumbo
for over 140000 people for free during his high energy concerts with his band--The
Ingredients">
<meta name="GENERATOR" content="Mozilla/4.73C-CCK-MCD {C-UDP; EBM-
APPLE} (Macintosh; U; PPC) [Netscape]">
<meta name="keywords" content="blues music hot sauce gumbo slide guitar Bill
Wharton datil pepper habanero Liquid Summer recipe contest Sauce Boss Jimmy
Buffett Parrothead Planet Gumbo Podcast Florida Blues">
<link rel="stylesheet" href="main.css" type="text/css" media=screen><title>Sauce
Boss</title>
</head>
```

(www.sauceboss.com)

Meta Tags

Meta tags are author-generated HTML commands that are placed in the head section of an HTML document. These tags help identify the content of the

Table 6.1	Source Code for Head Tags
Tag	**Description**
<!DOCTYPE>	Defines the document type; goes before the <html> start tag
<head>	Defines information about the document
<title>	Defines the document title
<base>	Defines a base URL for all the links on a page
<link>	Defines a resource reference
<meta>	Defines meta information (as in meta tags—see the text)

page and specify which search terms should be used to list the site on search engines. Popular meta tags can affect search engine rankings and are generally listed in the sections "Meta Keywords" and "Meta Description." Search engines often use the description meta tag and display it in the results. A meta tag can be generated automatically by the site www.submitcorner.com. The example tag shown here will let a search engine know to categorize the above web site under blues music, slide guitar, and by the artist's other endeavor, Louisiana-style cooking.

The web page title is also significant and should reflect the nature of the site. The <TITLE> tag is the caption that appears on the title bar of your browser and is the name on the clickable link listed in the search engine results. (Example: <title>Sauce Boss</title>.)

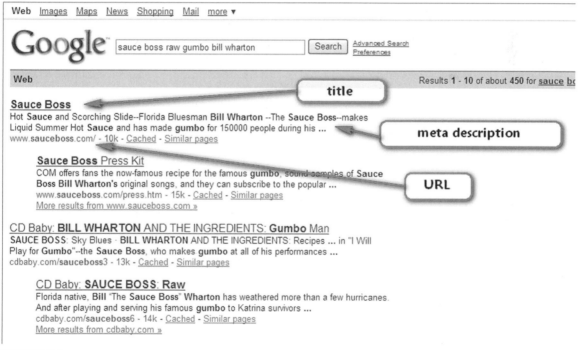

FIGURE 6.1
Google search results for Sauce Boss, including tags. (Courtesy of Bill Wharton.)

The table shown in Figure 6.2 is from www.makemyownwebpage.com and describes the various types of meta tags. The author describes meta tags as "where the search engines go to see what your priorities are."

FIGURE 6.2
Example meta tags (www.makemyown webpage.com) by permission.

META Tags

The <meta> tags are where the search engines go to see what your priorities are. A couple of other resources are also looking at meta tags for other content. You may have several Meta tags for a single web page.

Syntax

```
<META NAME="item-name" CONTENT="items">
<META HTTP-EQUIV="items" CONTENT="item-coding">
```

Modifiers

NAME="DESCRIPTION" CONTENT="*description*"
  This description is what most search engines tell the world about your web page.
NAME="KEYWORDS" CONTENT="*key, word, keyword*"
  Keywords help define your topic for what the search engines are looking for.
  Separate each keyword or phrase with a comma.
NAME="ROBOTS" CONTENT="*index,follow*"
  Tells the Search Engine Robots how to look at your site. The possible choices here
  are first *INDEX* or *NOINDEX*. This tells the search engines whether or not you
  want them to scan this page and include it in their indexes. Second term is
  FOLLOW or *NOFOLLOW*, which tells the search engines to follow the links on the
  page and index those or not. Not all search engines recognize this term.
NAME="RATING" CONTENT="*general*"
  This helps define the level of objectionable materials within this web page. Possible
  choices here are *GENERAL* (general audience... anybody), *MATURE* (Adult
  materials... may be objectionable), *RESTRICTED* (18 years old and up) and *14
  YEARS* (14 years old and up.
NAME="AUTHOR" CONTENT="*mtg@makemyownwebpage.com*"
  Lists the authors name and/or email address.
NAME="COPYRIGHT" CONTENT="*Copyright 2002*"
  Displays the copyright, trademark or any intellectual property information about the
  web page.
NAME="GENERATOR" CONTENT="*notepad*"
  Displays the publishing tool used to make my own web page.
HTTP-EQUIV="refresh" CONTENT="*0; URL=index.htm*"
  This allows you to forward the visitor to a new page as listed by the URL term. The
  timing (just before URL) is set to how many seconds to pause before forwarding.

Body Tags

The body of an HTML document uses tags to format text and graphics. There are numerous tags to denote font size, color, style, and other features, and there are others that denote paragraph formatting. Figure 6.3 indicates some of the body text formatting tags commonly used, including break, alignment, and justified.

```
<body>
This is line one followed by a break.<br>
And here is line two with a break.<br>
And a third and final line.<p> </p>


<p>
This is line one followed by a break.<br>
And here is line two followed by a paragraph tag.</p>
<p>And a third final line.</p>




<p align="left">Left Justified Text</p>
<p>Here is a small block of text, enough text to cause the sentence to wrap
  around and illustrate the effect of the "left" tag on word wrap.</p>




<p align="right">Right Justified Text</p>
<p align="right">Here is a small block of text, enough text to cause the
sentence to wrap and illustrate the effect of the "right" tag on word
wrap.</p>




<p align="center">Center Justified Text</p>
<p align="center">Here is a small block of text where the sentence wraps
and illustrates the effect of "center" on word wrap.</p>




<p align="justify">Justify Justified</p>
<p align="justify">Here is a small block of fully-justified text, enough to
wrap and illustrate the effect of "justify" on word wrap.</p>
<p>
```

This is line one followed by a break.
And here is line two with a break.
And a third and final line.

This is line one followed by a break.
And here is line two followed by a paragraph tag.

And a third final line.

Left Justified Text

Here is a small block of text, enough text to cause the sentence to wrap around and illustrate the effect of the "left" tag on word wrap.

Right Justified Text

Here is a small block of text, enough text to cause the sentence to wrap and illustrate the effect of the "right" tag on word wrap.

Center Justified Text

Here is a small block of text where the sentence wraps and illustrates the effect of "center" on word wrap.

Justify Justified

Here is a small block of fully-justified text, enough to wrap and illustrate the effect of "justify" on word wrap.

FIGURE 6.3
Example of page body formatting.

Other tags may denote font style and color. It is recommended that you use standard, commonly used fonts. If the web visitor's browser or computer is not set up for the fonts you have chosen for your body text, the substitution fonts may disrupt the formatting of the page and cause elements to shift out of place.

FONTS AND COLORS

HTML tags also indicate characteristics of the text appearance such as font size, color, and formatting.

In Figure 6.4, Arial and Times Roman fonts are used, along with underline, italics, bold, and font colors. The <i> tags indicate italics, <b> is for bold type, <u> is for underline, and the font tag can include, among other things, face, color, and size.

FIGURE 6.4
Example of font formatting.

```
<font face="Arial">Arial 12 point font</font><p><b><i>
<font face="Times New Roman" size="4">Times Roman 14 point type in bold
and italics</font></i></b></p>
<p><font face="Times New Roman" size="4">
This sentence contains <u>underlines</u>,
<i>italics,</i> <b>bold,</b> and
<font color="#0000FF">a color change to blue</font>
</b></font></p>
```

Arial 12 point font

Times Roman 14 point type in bold and italics

This sentence contains <u>underlines</u>, *italics,* **bold,** and a color change to blue

Table 6.2	Commonly Used HTML Formatting Commands	
Basic Text Commands		
 	Break	Same as carriage return.
<p>	Paragraph	Carriage return plus adds a blank line.
<b> </b>	Bold	Text between the two tags is displayed in bold.
<i> </i>	Italics	Text between the two tags is displayed in italics.
<u> </u>	Underline	Text between the two tags is displayed in underline.
<font size=?></font>	Font size	Sets size of font from 1 to 7.
<font color=?></font>	Font color	Sets font color, using name or hex value.
<font face=?></font>	Font typeface	Sets the typeface such as Times Roman, Arial, Tahoma, etc.
<font size="3" color="#FF000" face="arial">The actual text affected by the commands</font>		Example of font size, color and face.

Table 6.2	Commonly Used HTML Formatting Commands—cont'd	
Images		
<image src=url>	Insert an image	url would be the address where the image is found and generally ends with .jpeg or .gif.
Image subcommands	Width = 120	The picture width will be displayed as 120 pixels wide.
	Height = 60	The picture width will be displayed as 60 pixels high.
	Align = left	The picture will align to the left of text.
	ALT = "text"	Tells the browser what text to insert on the screen if the image is not available.
	Border = 4	The border around the image will be 4 pixels.
	Hspace = 4	The horizontal space between the image and surrounding text will be 4 pixels.
	Vspace = 4	The vertical space between the image and surrounding text.
Links		
<a href="mailto:EMAIL"></a>	E-mail link	Creates a mailto link that automatically opens up the user's default e-mail program and inserts the e-mail address.
<a href="newpage.html">Link to New Page</a>	Hyperlink to another site or location	The "Link to New Page" portion of the command is what is displayed on the screen as the hot link.

(Continued)

Table 6.2	Commonly Used HTML Formatting Commands—cont'd	
Color Commands		
Hex HTML colors are indicated by a six-digit series of letters and numbers but can also be indicated by a color name.		
bgcolor="#000000"	Background color	Used to tell the browser what color of background to use.
text="#ffffff"	Text color	The standard color to use on general text that follows.
link="#004000"	Link color	The standard color to use on unvisited links that follow.
vlink="#44aaff"	Color for visited links	
alink="#ff00ff"	Color for mouse-over	As the cursor passes over the link, this is the momentary color the link changes to while the user is hovering over the link with the cursor.
background="file.gif"	Background image	If you prefer to use an image as the background, this command specifies the image.

CGI SCRIPTS

Common Gateway Interface (CGI) scripts are defined as script files executed on a web server in response to a user request. They are commonly used to process data sent when a form filled in by a user is sent back to the web server. A CGI program is executable, and is basically the equivalent of letting visitors run a program on your system. The most common form found on artist web sites is the e-mail registration form that gathers e-mail addresses from visitors so that they may be added to the mailing list. The "Tell a friend" form is another example of a GCI script.

Forms

Forms are a common and popular feature on web sites. The e-mail signup and tell-a-friend scripts mentioned later are examples of simple forms. But forms are used for all kinds of purposes, from gathering information to processing e-commerce orders. Most high-end web development programs offer some type of features for creating forms. One of the important aspects of creating a form is determining where the information will go—either stored as a database on a server or sent as an e-mail to the appropriate person. The first step involves creating the form and specifying the action to be taken upon submission (where the data go). Then, the particular form fields can be created to elicit information from the web visitor. There are also companies online who provide form or survey features. Some start with a basic, free plan with many of the more advanced features reserved for the paid plans. These services have user-friendly web sites that allow even the novice to create a form, save the data, and analyze the results. The most popular of these is Survey Monkey. The company hosts the forms on its site. Another such service, Freedback, does not host the form but helps users create the form and then capture the HTML code to place on the Freedback web site. Survey Monkey stores the data—the filled-out forms—on its web site with convenient features for analyzing or downloading the data. Freedback sends the data to the e-mail address specified on the account.

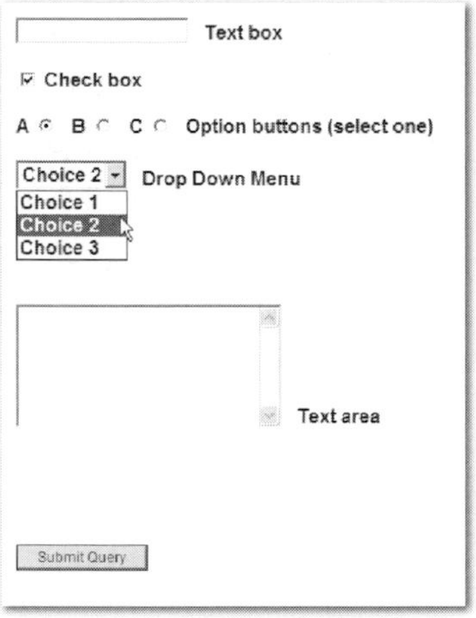

FIGURE 6.5
Example of forms.

Guestbook Script

Here is a simple CGI script for creating a form for signup for a mailing list:

```
<FORM ACTION="http://www.artistwebsite.com/mailinglist/subscribe.pl"
METHOD="post"><INPUT TYPE="hidden" NAME="account"
VALUE="hutchtom"><INPUT TYPE="hidden" NAME="body" VALUE="<BODY
BGCOLOR=white TEXT=black LINK=blue VLINK=darkblue>"><INPUT
TYPE="hidden" NAME="action" VALUE="subscribe"><TABLE BORDER="2"
CELLPADDING=0 CELLSPACING=0><TR><TD><TABLE BORDER="0"
CELLPADDING=2 CELLSPACING=0 BGCOLOR="#eeeeee"><TR
BGCOLOR="#cccccc"><TD><B>Your e-mail:</B></TD><TD><INPUT TYPE="text"
NAME="email"></TD></TR><TR><TD COLSPAN="2"><INPUT TYPE="submit"
VALUE="Join mailinglist!"></TD></TR>
</TABLE></TD></TR>
</TABLE>
</FORM>
```

Most commercial web entities want more information than just the person's e-mail address, so more elaborate scripts ask for additional information, including demographics. This script involves storing the information in a database to be accessed by the web master.

FIGURE 6.6
Join Now! form.

TELL-A-FRIEND SCRIPT

The tell-a-friend script is illustrated in Figure 6.7. As a result of automated spamming programs that are abused by unscrupulous web marketers, many such forms now include a component to verify that a human is filling out the form, instead of an automated software program.

```
<FORM ACTION="http://www.javascript.nu/cgi4free/tellafriend/tell.asp" METHOD="post">
<INPUT TYPE="hidden" NAME="yoururl" VALUE="http://www.hutchtom.com">

<TABLE BORDER="2" CELLPADDING=0 CELLSPACING=0>
<TR><TD><TABLE BORDER="0" CELLPADDING=2 CELLSPACING=0 BGCOLOR="#eeeeee">
<TR BGCOLOR="#cccccc"><TD COLSPAN="2"><B>Tell a friend about this site</B></TD></TR>
<TR><TD>Your name:</TD><TD><INPUT TYPE="text" NAME="name"></TD></TR>
<TR><TD>Your e-mail:</TD><TD><INPUT TYPE="text" NAME="from"></TD></TR>
<TR><TD>Friend's e-mail:</TD><TD><INPUT TYPE="text" NAME="to"></TD></TR>
<TR><TD COLSPAN="2"><INPUT TYPE="submit" VALUE="Tell a Friend!"></TD></TR>
</TABLE></TD></TR>
</TABLE>
</FORM>
```

The code above will result in this form:

Tell a friend about this site
Your name: [   ]
Your e-mail: [   ]
Friend's e-mail: [   ]
[Tell a Friend!]

FIGURE 6.7
Tell-a-friend form. (Courtesy of CGI4Free.com.)

JAVA AND JAVASCRIPT

Whereas CGI is for server-side programming, often referred to as back-end programming, JavaScript is used for client-side programming, often referred to as front-end programming (although it can also be used for server side programming). Server side programming runs on the host's server, whereas client side programming runs in the user's browser. Wikipedia has the following paragraph to explain the difference:

> Server-side scripting is a web server technology in which a user's request is fulfilled by running a script directly on the web server to generate dynamic HTML pages. It is usually used to provide interactive web sites that interface to databases or other data stores. This is different from client-side scripting where scripts are run by the viewing web browser, usually in JavaScript. The primary advantage to server-side scripting is the ability to highly customize the response based on the user's requirements, access rights, or queries into data stores.

According to Ibama Tmunotein in his article *Client-side and Server-side JavaScript,* "Server-side JavaScript is ideal for creating web applications that can be run on any platform, on any browser, and in any (programming) language."

Java is a programming language developed by Sun. NetScape responded by creating JavaScript. Microsoft then added its own version of JavaScript to Internet Explorer, called JScript (D. Smith, 1998). The difference between Java and JavaScript is that Java can stand on its own whereas JavaScript must be placed inside an HTML document to function. JavaScript is text that is fed into a browser that can read it and then is enacted by the browser. It can be modified on the fly. Java, on the other hand, creates a "standalone" application—the Java "applet" (a small application), which is a fully contained program. Java needs to be recompiled if it is modified, and then reinserted into the web page.

RESOURCES FOR SCRIPTS

Online HTML Code Generator, http://htmlcode.discoveryvip.com

Click & Go drop down list generator,
   www.webdevtips.co.0uk/webdevtips/codegen/clickgo.shtml

Meta tag and SEO generators, www.submitcorner.com

Guestbook and mailing list script generators, www.javascript.nu/cgi4free

HTML and CSS scripts, www.hypergurl.com/htmlscripts.html

Updated information available at www.WM4MB.com.

WEB WIDGETS

Web widgets are defined as a portable piece of code that an end user can install and execute within any separate HTML page. They often use DHTML, Adobe Flash, or JavaScript programming language and wrap it up in a nice user interface. Often these widgets are incorporated into social networking pages, blogs, and personal web sites, installed by the user. Not all widgets are compatible with all systems. In 2007, known as the year of the widget, the top social networking sites began to open up their platforms to widgets such as iLike, developed by third parties.

Marketers are creating new ways to use widgets to advertise and sell products. Here is how widgets such as iLike work. The widget company hosts a web site where the end user can construct the personalized parameters of the widget. The user-friendly interface allows the user to customize and personalize the widget, such as create a playlist, upload favorite photographs, and so forth. Then with a mouse click and a password, users upload the widget to their social networking page, which seamlessly and transparently adds the special feature to their social networking page or web page. This allows for content to be dynamic as the widget updates information based on the user's choices and activity—such as monitoring iTunes listening activity to create a playlist to be posted on a social networking site or whenever the user adds photos to the web site that provides the service and the widget.

FIGURE 6.8
OnTour widget. (Courtesy
PassAlong Networks
www.passalong.com.)

Fore example, the web site www.slide.com allows users to upload pictures and create slide shows that they can share with other web users. The slide.com site creates widgets that work with MySpace, Facebook, Bebo, Friendster, Orkut, Hi5, Tagged, Xanga, and the FunWall for Facebook users. Music marketers have starting using widgets to allow fans to add content to their social networking pages. One such company, PassAlong Networks, offers widgets with features that promote the artists the company is working with. These widgets can be found on the artists' web sites and allow fans to place this artist-featured content on their own sites.

CAPTCHA

In today's age of spamming, web site managers have adopted techniques to prevent automated programs from performing functions that are supposed to be performed by human visitors to the site, such as posting messages on a message board or sending e-mails. Typically, this is achieved through a CAPTCHA. CAPTCHA is a program that can tell whether its user is a human or a computer. It is a loosely contrived acronym meaning "completely automated public turing tests to tell computers and humans apart." CAPTCHAs are graphics presented with distorted text found at the bottom of web registration forms. Many web sites use CAPTCHAs to prevent abuse from "bots," or automated spamming programs. No computer program can read distorted text as well as humans can, so bots cannot enter sites protected by CAPTCHAs. Thus, e-mail accounts and

FIGURE 6.9
A typical graphic
verification image.

message boards can be protected through the use of CAPTCHA programs. It can also protect forms and online polls. The process involves installing a program that can generate and grade tests—in this case, an easy test of repeating the letters and numbers that appear in the distorted graphic. The concept was developed at Carnegie Mellon's CyLab, and the code is now offered for free at http://recapture.net.

FLASH

Flash, a popular authoring software developed by Macromedia and now owned by Adobe, is used to create graphic/animation programs with navigation interfaces, graphic illustrations, and simple interactivity in a resizable file format that is small enough to stream across a normal modem connection (although most Flash content over a "normal modem connection" is disturbingly slow to load). It is platform independent and gives web designers the ability to add bells and whistles to animate any web page. Wikipedia states, "Flash technology has become a popular method for adding animation and interactivity to web pages; Flash is commonly used to create animation, advertisements, various web page components, to integrate video into web pages, and more recently, to develop rich Internet applications" (wikipedia.com). The end user must have Adobe's Flash player, a free software download, installed to run the Flash programs. Adobe sells the software program to develop Flash programs to add to web sites.

In the article "What Is Flash, When and Why to Use It," author Stefan Mischook weighed the pros and cons of using Flash, emphasizing that it's not the best option for all applications, stating that Flash should be used "selectively to enhance an HTML-based site." One major advantage is that there are no compatibility issues with the various browsers and platforms on the market. Aycan Gulez, in his article "Top Five Reasons for Limiting Flash Use," stated that the Adobe Flash plug-in is installed in over 95% of web users' browsers, so just about all users can display Flash content. But, on the flip side, he emphasized the limitations of Flash including limited navigation ability (you may need to wait until it's finished loading to move on), slow load times, poor rendering of text, and the inability to print out or search through text. The general rule on Flash-heavy splash pages is to offer the web visitor the option to skip the Flash introduction.

There are other, less-expensive programs that can create the same effect, including mix-fx (www.mix-fx.com). CoffeeCup.com offers software that will create particular Flash effects. Glogster.com offers limited online templates from which you can develop a multimedia Flash presentation at no charge (www.glogster.com).

RSS FEEDS

Wikipedia describes *really simple syndication* (RSS) as

> a family of Web feed formats used to publish frequently updated content such as blog entries, news headlines or podcasts. An RSS document, which is called a "feed," "web feed," or "channel," contains either a summary of content from an associated web site or the full text. RSS makes it possible for people to keep up with their favorite web sites in an automated manner that's easier than checking them manually.

RSS was first invented by Netscape, which wanted a way to get news stories and information from other sites and have them automatically added to its site. The RSS feed starts with the XML command (Extensible Markup Language), then the RSS command, some commands for formatting, and the location (link) for the feed.

> We found early on the use of RSS feed to help market and promote our clients in a unique way. Using the system to push news, tour dates, and blogs and journals to other sites and subscribers, we have developed those efforts, while the Internet as a whole has adapted its use. With the inventions of widgets and sidebars for Windows, Macs and even more so Vista, we are able to take the RSS feeds we have been using for several years and hit a whole new target easier (the desktop of the consumer). We can now send snippets of code that will show the latest news and tour information right on their desktop or email box without requiring them to install something new.
>
> **Stephanie Orr-Buttrey, www.CountryWired.com**

The music business uses RSS feeds to keep fans informed of upcoming events or to provide updated content for fan-based web pages. This actually allows the "feeder" (the artist's web manager) to control content that appears on other web sites quickly and easily with one feed.

FIGURE 6.10
Illustration of how RSS feeds update fan sites.

TESTING ACROSS BROWSERS AND PLATFORMS

As illustrated in Chapter 5 on web design, a web layout that looks correct and normal on one computer may look completely different on another computer. The layout of a web site varies with the following characteristics:

- Type of browser (FireFox, Internet Explorer, Safari, etc.)
- Version of browser
- Browser settings
- Computer operating system
- Monitor resolution

In his article "Browser Compatibility Tutorial," Tom Dahm explained why a web site looks different depending on the preceding factors. He described the browser as "a translation device ... it takes a document written in the HTML language and translates it into a formatted Web page." The standards for HTML and browser compatibility are setup by the World Wide Web consortium (W3) that publishes these standards. But as HTML evolves, the older browsers, and even some of the newer ones, fail to keep up and support all the newest bells and whistles. Generally, the newer browser versions are more standardized than the older versions. That presents problems for web designers who want to use all the latest features, some of which are not supported for the portion of web visitors who are still using the older versions.

Font availability and size can be problematic. A web site may be created in a font style that is not available to many of the site's visitors. Fonts reside on the user's computer and are put into place based on the HTML instructions. If that font is not available on that computer, the browser substitutes another font, sometimes with grave consequences. Font size can cause problems also. Many browsers allow users to customize their default font size. Some users prefer to increase font size to reduce eyestrain. This may lead to a web page that is out of proportion, with text dominating the page. It is wise to use standardized fonts when creating a web page. If unique fonts are to be used, they should be converted to graphic files (preferably GIF files) so that consistency will be maintained across platforms.

Once the web site has been created and uploaded, it is wise to check out compatibility by either visiting the site on a variety of different computer setups or using one of the online services that simulate or capture different browser experiences. Other tips include the following:

- Don't build your web site entirely with Flash, and use a dedicated web design program instead of Microsoft Word or Microsoft Publisher.
- Check your web page HTML at http://validator.w3.org or www.anybrowser.com/validateit.html.
- Check your page's/site's CSS at http://jigsaw.w3.org/css-validator.
- Check your links at http://validator.w3.org/checklink or www.anybrowser.com/linkchecker.html.

- Check your images for accessibility issues at http://juicystudio.com/services/image.php.
- Check your content for readability at http://juicystudio.com/services/readability.php.
- Check to see if your CSS's text and background colors have sufficient contrast at www.accesskeys.org/tools/color-contrast.html.
- Check how your site is viewed on different systems at www.anybrowser.com/siteviewer.html, www.anybrowser.com/ScreenSizeTest.html, or www.browsercam.com/Default2.aspx.
- Check your page's performance and web page speed at www.websiteoptimization.com/services/analyze.
- Check several things at once with www.netmechanic.com/products/HTML_Toolbox_FreeSample.shtml.

GLOSSARY

Applet – A small Java program that is cross-platform compatible and can be embedded in the HTML of a web page. Web browsers, which are usually equipped with Java virtual machines, can run the applets to perform interactive graphics, games, and so on.

Browser – A software application used to locate and display web pages. Contemporary browsers are graphical browsers, meaning they can display graphics as well as text and can present multimedia information, including sound and video, though they require plug-ins for some formats.

CAPTCHA – A program that can tell whether its user is a human or a computer. The process involves installing a program that can generate and grade tests—in this case an easy test of repeating the letters and numbers that appear in the distorted graphic that humans can read and software programs cannot.

CGI scripts – Common Gateway Interface (CGI) scripts are defined as script files executed on a web server in response to a user request. Used for user-generated forms.

Client-side JavaScript (CSJS) – JavaScript that enables web pages and client browsers to be enhanced and manipulated.

Client-side programming – Occurs on the end-user side of a client-server system—these programs are executed by your browser (the client).

CSS – Short for cascading style sheets, a feature being added to HTML that gives both web site developers and users more control over how pages are displayed. With CSS, designers and users can create style sheets that define how different elements, such as headers and links, appear. These style sheets can then be applied to any web page.

Flash – A bandwidth-friendly and browser-independent animation technology. As long as different browsers are equipped with the necessary plug-ins, Flash animations will look the same. With Flash, users can draw their own animations or import other vector-based images (webopedia).

Hypertext markup language (HTML) – The predominant authoring language for the creation of web pages. HTML defines the structure and layout of a web document by using a variety of tags and attributes.

Java – A client-side programming language with a number of features that work well on the Web. Small Java applications are called Java applets and are downloaded from a web server and run on the user's computer by a Java-compatible web browser.

JavaScript – A server-side scripting language embedded in the HTML language of a web page that adds interactive functions to HTML pages. JavaScript is easier to use than Java, but it is not as powerful and deals mainly with the elements on the web page.

Java Virtual Machine (JVM) – An abstract computing machine, or virtual machine, is a platform-independent execution environment that converts Java byte code into machine language and executes it.

Meta tags – Author-generated HTML commands that are placed in the head section of an HTML document. These tags help identify the content of the page and specify which search terms should be used to list the site on search engines.

Plug-in – A hardware or software module that adds a specific feature or service to a larger system.

Server-side JavaScript (SSJS) – JavaScript that enables back-end access to databases, file systems, and servers.

Server-side scripting – Scripting that runs on the server side of a client-server system. CGI scripts are server-side applications because they run on the web server compared to programs that run in the user's browser.

Web widgets – A small application that can be ported to and run on different web pages by a simple modification of the web page's HTML.

XML – The Extensible Markup Language is a powerful tool for creating documents using structured information.

REFERENCES AND FURTHER READING

Burns, Joe. (2005, January 4). Java vs. JavaScript: HTML goodies, www.htmlgoodies.com/beyond/javascript/article.php/3470971.

Dahm, Tom. Browser compatibility tutorial, www.netmechanic.com/products/Browser-Tutorial.shtml.

Gulez, Aycan. (2001). Top five reasons for limiting Flash use, www.wowwebdesigns.com/power_guides/limiting_flash_use.php.

http://en.wikipedia.org/wiki/Captcha.

http://searchcio-midmarket.techtarget.com/sDefinition/0,,sid183_gci214563,00.html.

Kyrnin, Jennifer. What is RSS? http://webdesign.about.com/od/rss/a/what_is_rss.htm.

Mischook, Stefan. What is Flash, when and why to use it, www.killersites.com/articles/articles_FlashUse.htm.

Smith, Dori. (1998). What Is JavaScript? MacTech, volume 14, issue (5), www.mactech.com:16080/articles/mactech/Vol.14/14.05/WhatisJavaScript/?%2Findex.html.

Smith, Mike. A guide to HTML and CGI scripts, www.it.bton.ac.uk/~mas/mas/courses/html/html.html.

Swartz, John. (2007, November 27). Widgets make a big splash on the Net. USA Today.com. www.usatoday.com/money/industries/technology/2007-11-26-widgets_N.htm.

Tmunotein, Ibama Supreme. (2004, October 20). Client-side and server-side JavaScript. www.devarticles.com/c/a/JavaScript/Client-side-and-Server-side-JavaScript.

www.freedback.com.

www.humanverify.com.

www.iLike.com.

www.slide.com.

www.webopedia.com.

CHAPTER 7

Optimizing and Monitoring Your Web Site to Increase Visitation

Building a good web site is imperative for being successful on the Internet, but there are also certain modifications to a web site that can generate traffic to the site. Part of that involves adding elements to the web site that improve search engine rankings, called *search engine optimization* (SEO). Increasing traffic to the web site includes not only SEO, but also creating elements of the web site that will (1) bring in more visitors, (2) retain visitors longer, and (3) create more repeat traffic. Web surfers visit particular web sites because they are looking for something of interest. For music fans, information about music and the music itself play an important part, but it is also about the community and fun. Visitors need a reason to stay and a reason to return. This chapter discusses ways in which a web site can be modified to improve traffic to the site.

SEARCH ENGINE OPTIMIZATION

Search engine optimization is the practice of guiding the development or revamping of a web site so that it will naturally attract visitors by gaining top ranking on the major search engines for selected search terms and phrases. Chapter 11 discusses the process of submitting your web site to search engines to have them include your site in their directory. But you can improve site relevance, which helps determine how prominent your site is in the search engine results, by including a few extras in the site. The goal is to imagine what keywords your customers are likely to use when looking for sites such as yours.

Keywords

Search engines look for *keywords* when sorting and ranking sites. This includes the keywords found in meta tags (Chapters 6 and 11), but also text that appears in the first few paragraphs of the page to catalog "content-rich" sites. Because search engine placement is about beating the competition for the top slot in search results, the best place to start is by looking at the competition. Use one of the popular search engines and type in keywords that members of your target market would typically use to look for your artist or music. Don't be concerned about thinking of all relevant keywords at first—that will come later. As the results of the search are displayed, visit the first few listings to see if they are indeed targeting the same market and are considered competitors. You can actually use these web sites to improve your own. Look at the keywords they use by using your browser's "view source" function to look at their meta tags. Look at the text appearing on their home page and the titles used. Consider any words or phrases that you find on these sites that you may not currently be using or have thought about. Internet consultant Bruce Clay stated, "Proper Search Engine Optimization requires that you beat your competition, so knowing the keywords and criterion used by your competition is the most important first step" (Clay, 2007).

Several web companies specialize in SEO, and for a small expense, you can have the experts handle this. There are also web sites that will assist you in evaluating which keywords are most successful in generating traffic and top search engine placement. Some of these sites provide a basic service for no charge. There are two types of tools for managing keywords: (1) keyword generators and (2) keyword verifiers. Digital Point Management provides a keyword tracker tool that determines keyword popularity. Google Adwords provides a keyword tool to help Adword users determine the appropriate and most popular keyword phrases to use for their ads. By typing in a keyword statement or combination of keywords and clicking "get keyword ideas," the system displays results and indicates popularity for each suggested keyword phrase. Although these suggested phrases are for the benefit of AdWords users, the keywords generated by the program are also useful for inclusion in meta tags and in the text on the page. These and other programs help predict which keywords will be most successful for inclusion in your web site. Other software programs can test the effectiveness of your keywords already in use. By typing in your domain name and the keywords in question, the verification software can determine whether your web page shows up near the top of the search results for various search engines.

KEYWORD VERIFICATION TOOLS

Wordtracker.com

Google AdWords keyword tool,
     https://adwords.google.com/select/KeywordToolExternal

Digital Point Management keyword tool, www.digitalpoint.com/tools/suggestion

SEO Tools, www.seochat.com/seo-tools/keyword-suggestions-google

Keyword verification tool, www.marketleap.com/verify

Updated information available at www.WM4MB.com.

FIGURE 7.1
Keyword suggestion
tool from WordTracker
(www.wordtracker.com,
by permission).

Home / Keyword Suggestion Tool

FREE keyword suggestion tool

Keyword:
folk music

Adult Filter:
Off   Hit Me

**10 Great Reasons to
Subscribe to Wordtracker - Risk-Free!**

folk music

3,341 searches (top 100 only)	
Searches	**Keyword**
625	irish folk music
536	folk music
444	folk music lyrics
100	scottish folk music
87	folk music in new york city
79	places to audition for folk music
71	russian folk music
65	italian folk music
58	download turbo folk mp3 music
57	american folk music
52	history of irish folk music
49	german folk music
42	french folk music
42	old town school of folk music

Link Popularity

Another characteristic that search engines use to rank keyword results is *link popularity*; how many other sites think your site is important enough to link to it. The quantity (popularity) and the quality (relevance) of links to your site are used to determine ranking status. Link quality is defined as those from other sites with high page rankings for relevant search terms. Search engines use this information because they go by the assumption that the most important and relevant sites will have lots of other sites linking to it and also because it is hard for webmasters to fake or fool the search engine into giving a higher ranking than deserved. SearchEngineWatch.com states, "link analysis gives search engines a useful means of determining which pages are good for particular topics."

As outlined in Chapter 11, you can improve your link popularity by contacting webmasters at other relevant sites and asking them to place a link on their site that will lead visitors to your site. Use the search engines to find appropriate sites to request link placement. Type in the relevant keywords, and visit pages that appear at the top of the results. Ask those webmasters to add your link. Sometimes this is achieved through a link exchange. Chances are that a well-known music star is not going to agree to link to a site of an unknown artist, with the exception of major artists who are fans of the particular emerging artist. The best candidates for link exchange are similar artists and members of your fan base. Ask fans to link to your site, and perhaps offer a contest or incentive (a free download) for all those who comply.

When press releases or other written materials are disseminated, include several hot links to your artist's web site. As these articles are posted on web sites, e-zine sites, and blogs, the embedded links are spread virally and will appear in the electronically published version, thus increasing incoming link popularity.

Several free tools are available to measure link popularity. These tools search Google, AltaVista, MSN, HotBot, Lycos, and other search engines to determine how many pages are linking to your web page.

KEYWORD POPULARITY TOOLS

Market leap link popularity tool, www.marketleap.com/publinkpop

WideXL.com, www.widexl.com/remote/link-popularity

"TELL A FRIEND" AND "BOOKMARK THIS PAGE" SCRIPTS

Word of mouth is unquestionably the best form of marketing communication. It carries a sense of credibility lacking in most other forms of marketing. Generally the person who is giving the recommendation has some knowledge of the recipient's interests. The recipient is more inclined to pay attention to the recommendation because of the credibility of the source. Money can't buy this type of marketing, although many record labels spend much time and effort to create street teams to give the appearance of word of mouth or street credibility. Well-designed web sites make it easy for visitors to spread the word and pass along information about the site. The newest generation of social networking music-oriented sites even has software that allows a visitor to pluck e-mail addresses directly from their personal e-mail accounts to use when passing along a tip to visit a web site. One way that web sites make it easier to spread the word is through the "tell a friend" or "share this page" *JavaScript*—a bit of code embedded in one of the pages. Emarketer.com states, "A recent study showed that more than half (53%) of Internet users had visited web sites referred by friends or family members in the previous 30 days." Several web sites online will generate scripts for this feature. Some of the free versions add a viral message of their own, attached to the e-mail generated by the system and sent to the recipient of the referral. The "bookmark this page" feature is a bit of script to effortlessly add the web page to the user's list of favorites, using the page title found in the <head> tag as the name or listing for the page.

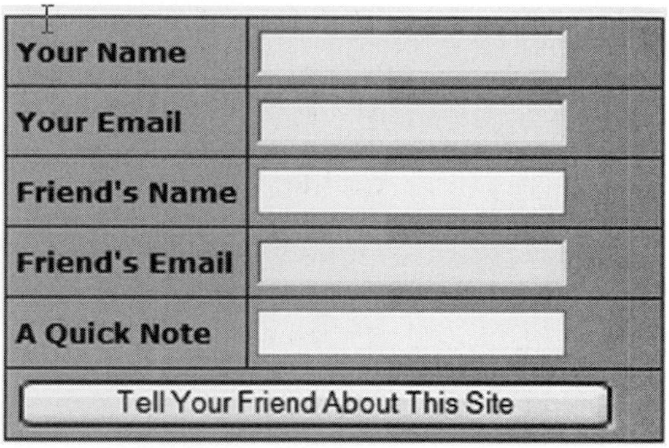

Your Name	
Your Email	
Friend's Name	
Friend's Email	
A Quick Note	
Tell Your Friend About This Site	

FIGURE 7.2
Tell-a-friend button.

TELL-A-FRIEND SCRIPTS

www.plus2net.com/php_tutorial/tell_friend.php
www.javascriptkit.com/script/script2/tellafriend.shtml

CONTESTS

Contests are a great way to generate traffic to the web site. They provide something of value to visitors and encourage them to return to the site. Contests should be run on a weekly or monthly basis, with one or more prizes awarded at the end of each contest period. The contest should be designed to maximize its benefit to the site owner. Visitors should be required to provide their e-mail address and other valuable marketing information to be entered in the contest. You can then use this information to create a mailing list, but be sure to alert visitors to this possibility and give them the opportunity to "opt out" of receiving future e-mail correspondence. Contests can also be designed to encourage visitors to further explore the web site. Some contests pose trivia questions, with the answers found on other pages on the site—like a scavenger hunt. Repeating the contest weekly or monthly, with a new entry required for each contest period, will encourage visitors to return to the site, perhaps bookmarking the site in their browser. Contests also increase the amount of recommendations that visitors send to their friends. They also create great linking opportunities. There are many sites that link to any site offering contests.

FIGURE 7.3
Example of web site contest for artist Trisha Yearwood. (Courtesy of CountryWired.com.)

General entry contests just require the visitor to "sign up for a drawing," but more creative contests can involve fans to a greater extent, thus making the contest more of a fun, interactive experience than just a game of chance. Fans can be asked to "pick the next" single after listening to several songs. Then a drawing can be held from among those who selected the song that winds up being the top pick. Another idea is to have fans send in their favorite digital photos from a recent performance of the artist. Contests have also been run where entrants create a new slogan, submit a song for consideration, create a *mashup*, come up with a new band name, or create a new band logo. If the purpose of the contest is to promote a new CD, it might not be best to use that CD as the contest prize. Fans may delay purchasing the album until the end of the contest. Offer something else of value: a T-shirt, an earlier release, or other swag.

BLOGS

Blogs or weblogs, have become popular on the Internet lately with the introduction of Web 2.0 and many sites and software programs that offer blogging opportunities. MarketingTerms.com defines a blog as

> a frequent, chronological publication of personal thoughts and Web links. A blog is often a mixture of what is happening in a person's life and what is happening on the Web, a kind of hybrid diary/guide site, although there are as many unique types of blogs as there are people.

Blogs are a good way to encourage repeat visits to your web site. Visitors know that with each visit to the artist's web site, they have a chance to read this fresh new material. The diary-like quality of blogs gives music fans the chance to feel closer to the artist, getting to know them better and establishing rapport. They bring a human element to an otherwise impersonal medium.

Blogs are also popular with search engines. The fresh, keyword-rich content of blogs is easy to find and catalog. Search engine spiders will crawl a site more frequently if it is updated regularly. Blogs frequently contain links to other information on the same site, luring the visitor deeper into the site. Blogs also commonly link to each other and these inbound links raise the popularity rating of the site. Blogging software and sites usually contain comments or feedback features, setting up a dialogue between readers and the blog author.

< Back　　　　　　　　　　　　　　　　　home | Unsubscribe | rss | add to blog group | sign out

Karg Boys

Last Updated:
Apr 6, 2008
Send Message
Instant Message
Email to a Friend
Unsubscribe
Invite to My Blog

Tuesday, June 26, 2007

Ramblings and Predictions

Well, well, well, here you are reading the Karg Boys' latest blog.....today you're reading from Richard's hand...I'm the one on the left side of the stage, if you've ever seen us live, and I'm on the right side of our myspace pic...I have the curls...

I am excited....we've been writing some great new songs lately...we've got some pretty underground recordings of them up to this point, and I'm not sure I'd distribute the current versions of these songs, but we are working them up and playing them live at this point....I can tell you that these new songs are amazing...you can see us live again on the 21st of July at the Bluebird Cafe's early show!

I'm very pleased to be a songwriter, mostly b/c I know that it is one of the 3 things in life that I'm really good at...and I may not be the best, but I'm good and I know I'm getting better...Andy's great too; not to leave him out...but I definitely feel like we are going to crank out some of the biggest songs of all time...KEEP AN EYE OUT!

1:36 AM - 3 Comments - 6 Kudos - Add Comment - 〇 SHARE

FIGURE 7.4
Artist blog entry. (Courtesy of the Karg Boys.)

All artists and musicians should maintain their own blog, creating entries when the muse strikes, but on a regular or frequent basis. Artists generally blog about experiences on the road or in the studio, and the fans enjoy the behind-the-scenes aspect of the artist blog. Artists have been known to pull out the laptop on the bus right after the event and write their latest blog entry. The instant publication of blogs makes it attractive to readers. Bloggers also use cell phone cameras to enrich the blog entries with photos. Software companies and web sites make it easy for bloggers who may not otherwise know how to post material on the Internet, and social networking sites generally provide easy-to-use blogging opportunities. Other sites, such as Blogger.com, offer services either for free or for a small fee. These blogs can then be embedded into the artist's site and fed to other web sites through RSS (see Chapter 6).

VISITOR REGISTRATION

The most valuable piece of information that can be obtained from consumers who visit the web site is their e-mail address and permission to add them to the "list." A list of e-mail addresses, and permission to contact the addressees, is necessary in this new age of *spamming*. Spamming is the activity of sending out unsolicited commercial e-mails. It is the online equivalent of telemarketing (more on spamming later). An effective, up-to-date e-mail list is a valuable marketing tool and allows for e-newsletters to be sent to fans who have shown enough interest to sign up. Web sites should post their privacy policies to avoid any confusion or legal complications if visitors end up on a mailing list.

When recruiting visitors to sign up, it is more attractive to present this as either a guest book or free membership to the artist's fan club, rather than just signing up to receive e-mails. Visitors can be enticed into signing up to gain "membership" to restricted areas of the web site that may contain free music downloads and allow fans to post messages on the site's message board. It is also a way for webmasters to monitor the site for inappropriate message posting and restrict access of repeat offenders. Members can also receive benefits such as priority in purchasing concert ticket and prerelease music purchases.

Usually, web sites include a visitor sign-in, registration, or "join now" button leading to a short online form. When visitors are asked to sign a guest book or register to enter the site, there are two strategies for adding these visitors to the e-mail list: "opt-in" and "opt-out." With opt-in, the visitor selects a blank check box to be added to the e-mail list. With opt-out, the button default is in the checked position and the visitor must uncheck the box to avoid being included in the e-mail list. "Opt-in" means that visitors choose to join a site's mailing list—one that is generally aimed at notifying the visitor of new developments. Some marketers set the default setting to the "opt-out" approach by automatically checking the "Yes, sign me up" box. By default, visitors overlook the box, thus "giving permission" to contact them. Marketers advise to use the opt-in method only. This avoids annoying visitors who did not intend to sign up for e-mails and simply overlooked the "uncheck" function. It also reduces the amount of follow-up work that the webmaster must do to honor unsubscribe requests.

FIGURE 7.5
Example sign-up form.

The Client Help Desk (www.clienthelpdesk.com) reports that almost 70% of Internet users say they unknowingly signed up for e-mail distribution lists. Almost 75% of those who received unsolicited e-mail took action to be removed from the sender's list. With opt-outs, the web site can claim a larger number of subscribers (the willing and the unknowing), whereas with opt-ins, the site can get a better understanding of how many people want to receive the e-mails or e-newsletters sent out from the list.

FAN-GENERATED CONTENT

In this era of Web 2.0, the power of creating content for the Internet has been turned over to the users, with sites like Wikipedia and imeem relying on the visitors to add meaningful content that attracts other visitors to the site. The Internet is full of such sites, and site visitors do not hesitate to pitch in their contributions. These creative participants are likely to encourage their friends to visit any sites where they have posted content—giving a viral aspect to the endeavor. So how can artist web sites take advantage of Web 2.0 user-generated features? By allowing fans to post messages for other fans and by creating a photo gallery whereby fans can post their digital photos or videos of themselves in activities related to the artist and their music—perhaps concert photos or photos and videos of them dancing to or singing karaoke versions of the artist's music. The Canadian music television channel MuchMusic created a program called "Show Me Yours."

> SHOW ME YOURS makes fans part of the show by allowing them to upload videos and photos from their mobile phone, desktop, or straight from their webcam to muchmusic.com, with the best clips airing on TV. Once uploaded, this content becomes viewable on muchmusic.com. With SHOW ME YOURS, anyone in Canada can be the star of their favourite program, and Canadians can see what other MuchMusic fans from across the country are posting.

> **(MuchMusic Press Release, November 2006)**

MONITORING WEB TRAFFIC

An important aspect of optimizing a web site involves monitoring traffic to the site. Web traffic refers to the number of visitors to your web site and the number of pages visited. Oftentimes, it is measured to determine the popularity of a web site and its individual pages and elements. By including a bit of programming code on each page of the web site, the webmaster can learn a lot about the visitors to the site. This helps the webmaster and other marketing professionals understand which elements of a web site are considered valuable to its visitors and which are not.

WHAT TO MEASURE

Some of the most important factors that are measured include the following:

1. *The number of visitors.* This is represented by the number of different people who access your web site over a period of time. From this information, you can determine which times are most popular for visitors. You can determine if your traffic is influenced by any marketing campaigns that may be unfolding, the impact of promotional materials such as e-mail blasts, and the impact of advertising. For example, you send out an e-mail blast to members of the fan club announcing a new tour schedule and notice a jump in the number of visitors to the site and the tour schedule page for the next couple of days.

2. *Whether these visitors are new or returning.* The effects of advertising and other marketing efforts to expand the market can be measured by observing the number of new visitors to the site. The number of

returning visitors indicates the success level of efforts designed to bring visitors back to the site, such as blogs, new material regularly posted to the site, and so on.

3. *The number of page views.* This is a measurement of how many pages each visitor looks at on the site. If the ratio is high, meaning that each visitor on average visits a fair number of pages, that is an indication of the "stickiness" of the site. Stickiness means that the site is so compelling that visitors are inclined to stick around and visit other sections. However, this could also indicate that they are not finding what they are looking for, so they keep going on to the next page hoping to find what they need. Determining which of these two factors is in play is covered by the next measurement.

4. *Time spent per page.* If visitors are spending a lot of time on particular pages, one could conclude that these pages contain something of interest to the visitor. If other pages are glossed over quickly, then perhaps they are not as meaningful to the visitor or the visitor has not yet found what they are looking for. If certain pages don't get much traffic, or visitors tend to spend little time on them, they should be reviewed to determine if the level of interest is appropriate (it may be a page designed for a subsection of visitors, such as the media) or whether the page should be revamped or combined with another page.

5. *Time spent on the site.* Visitors who spend a long time on the site are probably the most dedicated customers or fans, especially if they are returning visitors. The average amount of time spent on the site indicates the worthiness of the site in providing something of interest.

6. *Date and time.* It is helpful to know the most popular viewing times and days to plan when updates will be made to the site and if traffic is seasonal.

7. *Where visitors reside.* This information is not always accurate, as some visitors may use an Internet service provider (ISP) that reflects the location of the main servers instead of the visitor's hometown. But for most systems, country of origin and city are listed in the visitor statistics. You can determine if there is more activity on the web site coming from areas where the artist is touring. Then by combining that with information on page hits, you can determine how important or useful the tour information page is to visitors.

8. *Where visitors are coming from and which page they enter the site through.* This information can help you to determine which outside URLs are providing most of the traffic, whether it's other sites that link to yours, search engine traffic, or direct request (the user types in your domain name).

9. *Exit page.* Which page do visitors commonly view last before leaving your site? Sometimes the page content will help determine the reason people leave the site: they found what they were looking for, they didn't find what they wanted, you directed them elsewhere, or they made the purchase.

10. *The technology that visitors use.* This function indicates the resolution of the monitor, connection type, browser type and operating system of each visitor. It is helpful in determining whether users have the technology to handle the latest bells and whistles before deciding to add those features to the site.

How to Use That Information

Building an effective web site involves more than just aesthetics and design. The site has to offer some value to the visitor. Decisions on other aspects of marketing rely on input from customer feedback forms, surveys, and other devices. Often this requires effort on the part of the consumer to provide this valuable information to market research experts. One of the great advantages of the Internet is that it offers marketing analysts a rich body of marketing information based on where web visitors go, what they click on, and how long they engage with the marketing message. In an article "Five Reasons to Track Web Site Traffic," author Monte Enbysk pointed out that too many "small businesses build web sites, invest time in online marketing campaigns and then devote little or no effort to analyzing the return on their investment." He stated that *web analytics* tools can help the marketer in the following ways:

1. *Evaluate the effectiveness of marketing efforts.* You can see the results of each aspect of promotion and how it affects traffic to the site. You can find out what the keywords are that your customers use to find you.
2. *Figure out where your traffic is coming from.* This is generally known as the *referrer* function on web statistics programs. This function will let you know if your search engine optimization is working, or if most of your visitors are coming from some other source that you can devote more of your marketing efforts toward. You can determine if your advertising is working.
3. *Learn what your users like and don't like about your web site.* Find out if it's time to replace or take down those underperforming pages where visitors tend to bail out. You can assess modifications of an underperforming page by changes in visitor activity.
4. *Learn of defects in your site.* By analyzing what computer systems most of your visitors are using, you can determine if they are getting the benefit of your site design or if design elements are displaced because of incompatible systems and browsers.
5. *Get to know your customers.* After studying the data coming in and making adjustments to the site, you can learn what your visitors like and what they respond to. Tracking them can tell you what they are looking for when they visit your site.

WHERE TO FIND ANALYTICS TOOLS AND HOW TO APPLY THEM

The most basic tool first to appear on the scene is the simple hit counter, consisting of a bit of programming language that would place a component on the bottom of each page that counted the number of visitors. Marketing experts advise against the use of the simple hit counter, stating that this information should be kept private for use by the site webmaster and marketing team. No one is impressed with coming across a web site and finding out you are the "seventeenth visitor to this site"—period. The information is also limited compared to what is available on the market now.

Web analytics is defined as the use of data collected from a web site to determine which aspects of the web site work toward the business objectives. Many services on the Internet offer web analytics features. After asking a series of questions about how you want to track and compile information, the service will create the code to be inserted into every page of the web site. The code helps the service track activity on the site. The webmaster logs in to the service to view and download the statistics that the system has gathered.

FIGURE 7.6
Visible web counter.

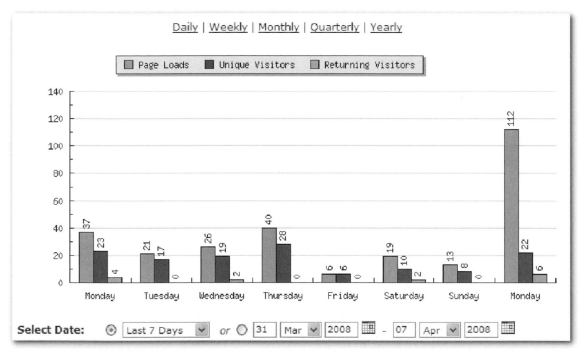

FIGURE 7.7
Visitor statistics.

GOOGLE ANALYTICS

In early 2005, Google bought the web analytics company Urchin. Later that year, Google revealed its repacked version of Urchin for free: Google Analytics. Google Analytics offers tracking of web visitors and provides the usual tracking statistics plus keyword reports and the ability to measure the effectiveness of AdWords programs. The process requires web owners to insert a small piece of Java code into the head tags of their pages. The statistical output can be viewed on the Google site with the assistance of a dashboard: customizable collection of report summaries.

STATISTICS AND WEB ANALYTICS TOOLS

www.OneStatFree.com provides a free, simple stat counter service and more advanced fee-based services.

www.Alexa.com monitors web traffic of people who use its toolbar. The site provides lists of most popular sites.

Nielsen Netratings is a commercial web monitoring service used by large sites.

www.trafficestimate.com is a free tool to help estimate the volume of traffic that a web site gets.

www.statcounter.com is a service that offers both free and paid statistics services.

www.iwebtrack.com is another service with free and paid versions of the service.

http://Analytics.google.com (see the accompanying box)

Updated information available at www.WM4MB.com.

COMPILING FAQS

What are FAQs?
FAQs, the abbreviation for Frequently Asked Questions, are a collection of the most commonly asked questions with the answers provided. This section may help a webmaster from having to personally answer e-mails asking the same questions time and again. It is also beneficial for placing information that will only be of interest to particular visitors.

How is FAQ pronounced?
Since the concept has its origins in text communication on the Internet, the pronunciation varies; it is pronounced as separate letters F.A.Q. or as "fak."

What does Wikipedia say about FAQs?
"Today 'FAQ' is more frequently used to refer to the list, and a text consisting of questions and their answers is often called a FAQ regardless of whether the questions are actually frequently asked (if asked at all). This is done to capitalize on the fact that the concept of a FAQ has become fairly familiar online." (Wikipedia http://en.wikipedia.org/wiki/FAQ)

What are the benefits of having FAQs on a web site?
Webmasters can learn a lot about what materials are missing or are obscure on the web site based upon visitor feedback and questions. From this, the list of FAQs is compiled and constantly revised to provide better service to the customer or site visitor.

Are FAQs commonly used on music web sites?
While it is not common to see a FAQ page on an artist web site, they are more common for record label sites. Some label sites have a FAQ page to let artists know how to submit material, or how candidates can apply for jobs or internships. A FAQ page is important whenever a contest is being held, to clarify entry policies and contest rules. Artists can also present an interview in FAQ form rather than standard interview form.

What are some example FAQs found on music web sites?
- How do I get in touch with the artists or set up an interview?
- How can I get fan club info?
- Can I use music and photos of Sony Music artists on my web site?
- Can I get permission to use lyrics and/or sheet music for a Sony Music artist's song?
- Where can I find sheet music?

From an archived Sony Music web site.

CONCLUSION

Building an attractive, enticing web site is very important. Adding these other elements will draw traffic to the site and encourage visitors to engage in word-of-mouth marketing. Search engine placement is an important aspect of helping people find your site. And once there, the elements of giveaways, contests, blogs, visitor registration, and fan-generated content encourage the visitor to return and tell others about your web site. Monitoring traffic to the site can be important for evaluating the effectiveness of each web page and for guiding the webmaster when making changes and upgrades.

GLOSSARY

Blog – (Weblog) a frequent, chronological publication or journal of personal thoughts and web links posted on the web.

FAQs – The abbreviation for Frequently Asked Questions, they are a collection of the most commonly asked questions with the answers provided.

Hot links – A link that takes the web browser to another place upon clicking the link.

JavaScript – A scripting language developed by Netscape and used to create interactive web sites.

Keyword – A word (or phrase) that a search engine uses in its hunt for relevant web pages.

Mashup – A mixture of content or elements

SEO (search engine optimization) – Various techniques that seek to improve the ranking of a web site in search engine results.

Spam – Flooding the Internet with many copies of the same message in an attempt to force the message on people who would not otherwise choose to receive it. Most spam is commercial advertising.

Spider – A program that automatically visits and catalogs web pages. Spiders are used to feed pages to search engines. Marketing companies also use them to gather information. The program is called a spider because it crawls over the Web. Another term for this type of program is webcrawler.

Web analytics – The use of data collected from a web site to determine which aspects of the web site work toward the business objectives.

REFERENCES AND FURTHER READING

Belba, Michael. Online contests draw website traffic, www.frugalmarketing.com/dtb/belbacontests.shtml.

Clay, Bruce. (2007). Search engine optimization (SEO), www.bruceclay.com/web_rank.htm.

Enbysk, Monte. Five reasons to track web site traffic, www.ecube.ca/5reasonstotrackWebsitetraffic.pdf.

Hurlbert, Wayne. (2004, July 13). Blogs as a website promotional tool, www.globalprblogweek.com/archives/blogs_as_a_website_p.php.

Hurlbert, Wayne. (2004, September 15). Blogs as excellent public relations tools, www.seochat.com/c/a/Website-Promotion-Help/Blogs-as-Excellent-Public-Relations-Tools.

Odden, Lee. (1/15/2007). SEO benefits from blogs, www.toprankblog.com/2007/01/seo-benefits-from-blogs.

SHOW ME YOURS on Muchmusic.com with new fan-generated content. (2006, November). Press release.

Sullivan, Danny. (2007, March 15). How search engines rank web pages, www.searchenginewatch.com/2167961/print.

Sullivan, Danny. (2007, March 15). Search engine placement tips, www.searchenginewatch.com/2168021/print.

www.searchenginewatch.com.

www.webanalyticsassociation.org.

Y-Times publication. (2007). Are you using Google Analytics? Here's why you should be, www.ytimes.info/areyouusgoan.html.

CHAPTER 8
Audio and Video for Your Web Site

Today, no music-related web site is complete without offering some music, either for download or streamed to the user's computer. In his book *How to Promote Your Music Successfully on the Internet*, author David Nevue stated, "if you want people to buy your music online, you've got to give them a sample of the goods." The question becomes how much audio? Should you provide samples of songs, entire songs, a few songs from each album? Nevue stated that some artists may not feel comfortable giving away their product, whereas others, including some major acts like Radiohead, are comfortable giving away their music if it increases their fan base and they can make up the money on concert tickets. The industry standard for online e-tailers is to offer 30-second samples of some or all songs on an album, but Nevue believes that song samples should be longer, up to two minutes, "enough for your potential buyer to get into the groove of your music" (Nevue, p. 29).

AUDIO FILES ON THE INTERNET

The creation of the *MP3* compression format for audio opened up the possibility for the first time of transferring music over the Internet. MP3 stands for MPEG (Motion Pictures Expert Group) Audio Layer III, and it is a standard for audio compression that makes any music file smaller with little loss of sound quality (although that's debatable). Without this compression technique, one second of CD quality sound requires 1.4 million bits of data, so the average song in *WAV* format is extremely large, averaging around 50 MB in file size. WAV is short for Waveform audio format, a Microsoft and IBM audio file format standard for storing audio on PCs. With MP3 compression, the file becomes one-twelfth its original size, at about 3 to 4 megabytes, depending on the bitrate.

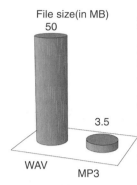

File size(in MB)

50

3.5

WAV

MP3

FIGURE 8.1
Comparison of file size
for WAV file and MP3.

MP3 was first introduced to consumers in 1997 with the launch of the popular Winamp player for computers and Diamond Multimedia's Rio portable MP3 player. The Recording Industry Association of America (RIAA) brought suit against Diamond, claiming that its Rio player did not fall under the protection of the *Audio Home Recording Act of 1992* because it did not employ a *Serial Copyright Management System* to protect the audio files from being exploited. The court rejected the argument, saying that because the obvious purpose of the player is for personal use, allowing consumers who owned a copy of the music to make a copy for their portable device constitutes fair use under copyright law.

MUSIC AND YOUR WEB SITE

A major issue for web site developers is whether to offer music as downloads or streaming. Do you want visitors to be able to click on a song and start listening immediately (streaming), or do you want them to be able to download, store, and own your song or sample on their computer (or portable device) so they can listen at their convenience? A comparison of *streaming* and downloading will be discussed next, followed by an overview of the different types of formats for audio files with and without digital rights management technology.

Streaming Audio

PC Magazine describes streaming audio as "a one-way audio transmission over a data network. It is widely used on the Web to deliver audio on demand or an audio broadcast (Internet radio)." With streaming, the audio file is not transferred to the user's computer but portions are stored in the computer's buffer so that the content will play normally, without interruptions.[1] (The other option is to allow users to download files to play at their convenience. This may lead to file sharing and does not provide the level of protection found with streaming.) The Information Technology web site at Cornell University describes streaming audio this way:

> When audio or video is streamed, a small buffer space is created on the user's computer, and data starts downloading into it. As soon as the buffer is full (usually 10–30 seconds), the file starts to play. As the file plays, it uses up information in the buffer, but while it is playing, more data is being downloaded. As long as the data can be downloaded as fast as it is used up in playback, the file will play smoothly.
>
> **http://atc.cit.cornell.edu/course/streaming/index.cfm**

According to David Nevue, you need to create an additional file called a "metafile" in order to stream MP3 files from your web site. This is done by creating a bit

---

[1] Depending upon the speed of the Internet connection. Streaming audio performs very poorly on dial-up systems. The portion in the buffer will play fine, but there will be significant delays as the player catches up to the buffer. The player will eat up data much faster than the buffer can refresh it.

of HTML language with a link to that MP3 file and saving that language in a file with the extension .m3u. When the user clicks on the link with the m3u extension, the default Internet audio player should open the original MP3 file and begin streaming it.

According to John Haring of Nashville Independent Music:

> This also can be accomplished in a cleaner manner by embedding a server-based player into the web site that will pop up when called. This is usually done with custom Flash code, although there are a number of open source players available to embed into your web site. This way, native mp3 files can be played without having to convert them and without the user having to have a player installed.

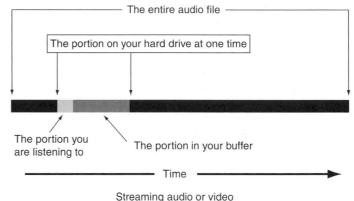

FIGURE 8.2
Illustration of streaming audio file.

You can also use Windows Media Encoder to create streaming audio and video in the *Windows Media Audio* (WMA) format (www.microsoft.com/windows/windowsmedia/forpros/encoder/default.mspx).

CREATING A MUSIC FILE FOR DOWNLOADING

For downloading, it is necessary to offer the audio file in one of the more commonly used compression standards. Setting up a downloadable file is much easier than setting up streaming—just set up a link to the file on the server and when web visitors click on the link, they will be able to download the song. Add all music files to a folder on the server and set up a link to each one. As visitors click on the link, they will be prompted to download the file. The disadvantages of this method include the fact that the visitor cannot begin to listen to the song until it has finished downloading. Also, there is less protection of the music if it is given away to visitors to download to their own computers; there is the potential for file sharing. Artists often reserve the free downloads for acoustic or live versions of songs rather than give away their primary product. In fact, giving away a "bonus track" can be a great promotional tool and generate traffic to the web site.

Audio Compression Formats for Downloading

The MP3 format is the most popular and most flexible audio compression format, supported by most platforms including the iPod. Only the older Sony standalone players fail to support it. As for computer software, Winamp, iTunes and Windows Media Player support the MP3 format. On the other hand, MP3 is the lowest-quality choice among the common formats, and, until recently, almost none of the legal music download services used it because of the lack of copy protection.

Advanced Audio Coding (AAC) is iTunes' compression of choice because of the sound quality and the copy protection incorporated into their files. iTunes and other major download retailers adopted *digital rights management* (DRM) early on to copy-protect downloads so they could not be swapped in peer-to-peer networks. This was the only way the download retailers could convince the major record labels to license their products for downloading. AAC is in either MPEG2 Advanced Audio Coding or MPEG4 Advanced Audio Coding. MPEG2 AAC can produce better audio quality than MP3 using less physical space for the files. MPEG4 AAC can produce even better quality and smaller files than MPEG2 AAC. In 2007, EMI and iTunes announced they would be offering a DRM-free version of EMI's catalog available on iTunes in the AAC format for the premium price of $1.29 per download. AAC files offer better sound quality and are around 30% smaller than the MP3 equivalent.

Windows Media Audio (WMA) is another format widely used by many online retailers, but it does not offer much flexibility. It outperforms MP3 in terms of quality and compression, particularly at lower bitrates. Consequently, WMA is probably the format of choice for streaming at low bandwidths. WalMart originally offered DRM-protected WMA format but added DRM-free MP3 format files in the third quarter of 2007.

Regardless of the format, any songs offered for download should be appropriately labeled with the song title, album title, artist, running time, and other important information. Be sure to include this information in any files created for download. Also, it may be wise to enter your track information in the universal Gracenote (CDDB) database. This is a web site that connects to audio software programs on computers and provides those computers with song identification data so that when the user loads up a CD in the computer, perhaps to rip the songs to a portable device, the software can download the track information from Gracenote. Any computer audio software program that works with Gracenote (such as iTunes or WinAmp) will enable you to enter in track information for your own CD and upload it to the system. Then when fans purchase your CD and play it in their computers, the correct track information will be available for them on their computer.

> The Gracenote Media Recognition Service is an Internet-based service that we license to software and hardware developers for use in their CD players, CD burners, MP3 players and encoders, catalogers, jukeboxes, cell phones, car audio systems, and home media center applications (among others). The service allows these developers to display artist, title, tracklists, and other music-related information automatically and instantly in their applications.
>
> For example, when you insert a music CD in your computer, the software player application on your computer uses our service to first identify the CD, and then display the artist, title, tracklist, and other information. Most commercial music CDs do not contain any of this information on the CD itself.
>
> **Gracenote.com FAQ**

PROVIDING MUSIC SAMPLES

One safe way to preserve the value of an artist's music is to provide 30-second samples instead of letting visitors download or listen to the entire song. No one wants to make and distribute copies of 30-second segments of a song. And the sample gives potential customers an idea of whether they might like the song or not. If the artist's music is featured for sale on one of the major online download services, they usually provide 30-second samples, and it may not be necessary to create them for the artist's site. Visitors can be redirected to one of those e-tailers to preview the music.

To provide samples on the artist's web site, they must first be edited from the full song. The idea is to select a sample that best represents the aura of the song, not simply start at the song's introduction and take the first 30 seconds. One general rule to follow is to capture the end portion of a verse and most of the first

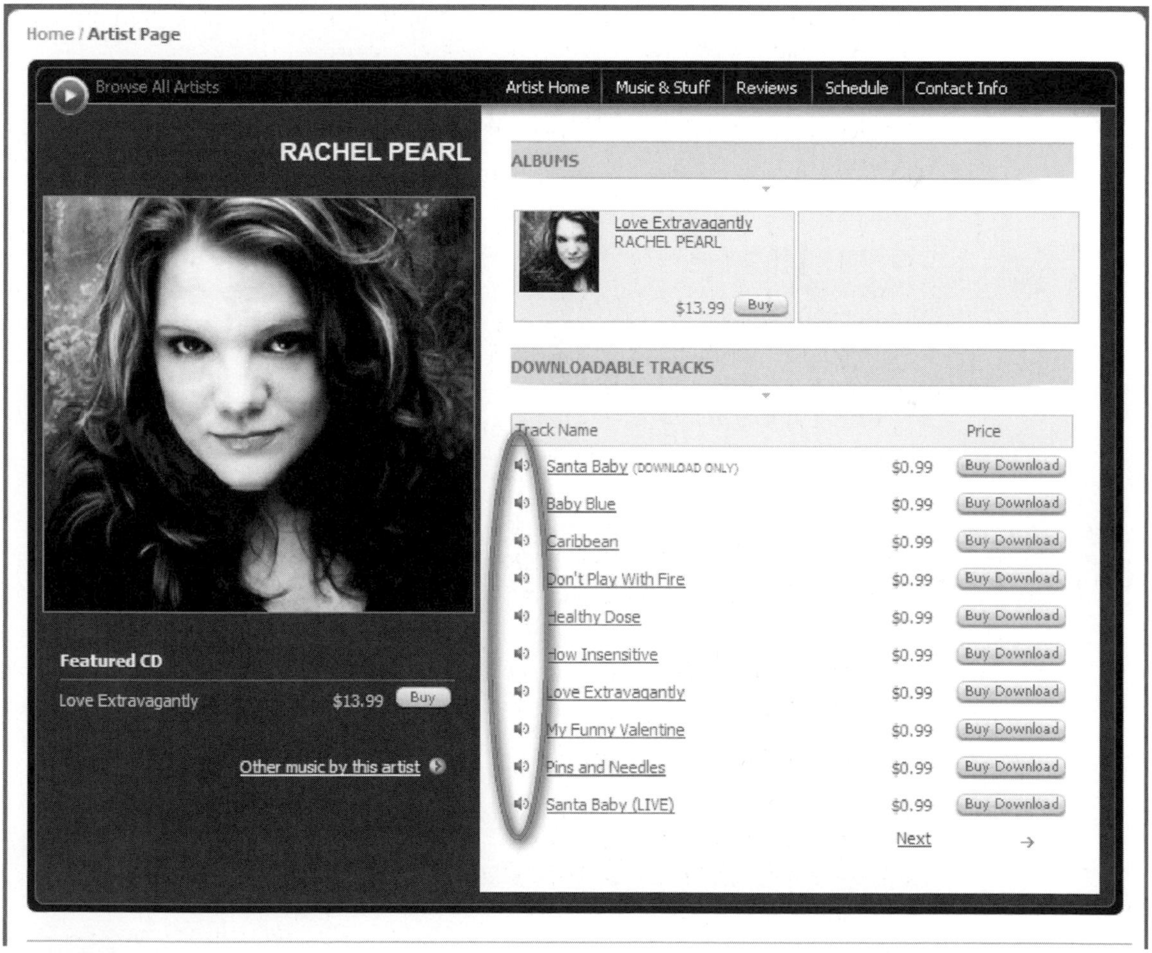

FIGURE 8.3
Audio samples indicated with the speaker icon. (Courtesy Nashville Independent Music, www.nashvilleim.com.)

chorus. At Nashville Independent Music, John Haring stated, "We've found that offering 45 seconds of a song starting from the 20 second point forward captures most of a verse and chorus. We use this standard when creating clips in our automated process for nashvilleim.com. We also automatically create a one second fade-in and a four second fade-out for better listenability."

Creating Music Samples from Songs

The process of creating samples can be accomplished using any music editing software such as ProTools, Cakewalk, or one of the less expensive audio editing programs available on the Internet such as Audacity or Gold Wave. From the songs selected for sampling, simply open the song in an editing program and listen to various 30-second sections until you have found a section that best represents the overall song. Then follow these simple steps:

In most audio editing programs, the running time is listed either at the bottom or the top of the song file. The graphic representation is amplitude modulation, with loud parts of the song showing up with large bars and quieter section showing shorter bars. With the highlighting tool, you can select 30 seconds and preview it to determine its suitability.

Once a 30-second sample has been selected that is a good representation of the song, highlight it and copy it to a new file for further editing.

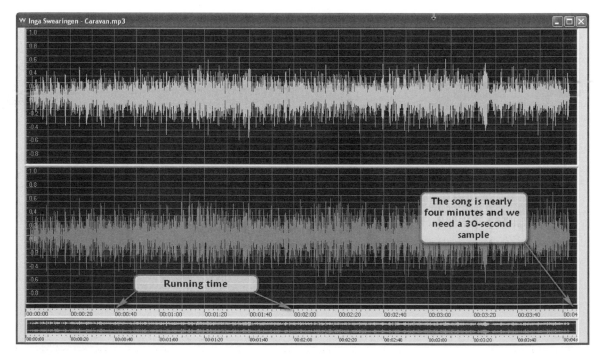

FIGURE 8.4
Graphical representation of song. (Permission from Goldwave.)

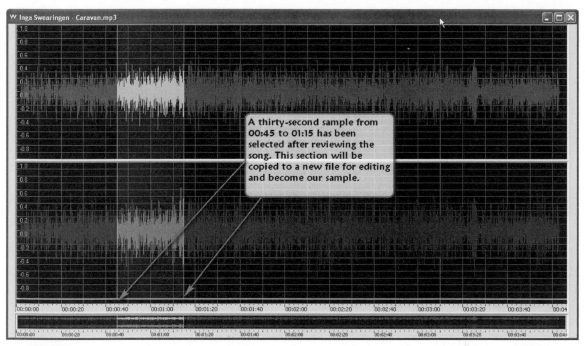

FIGURE 8.5
Highlighted 30-second song sample. (Permission from Goldwave.)

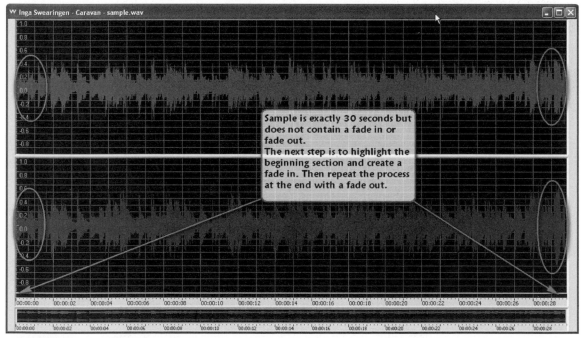

FIGURE 8.6
Sample expanded in a new file. (Permission from Goldwave.)

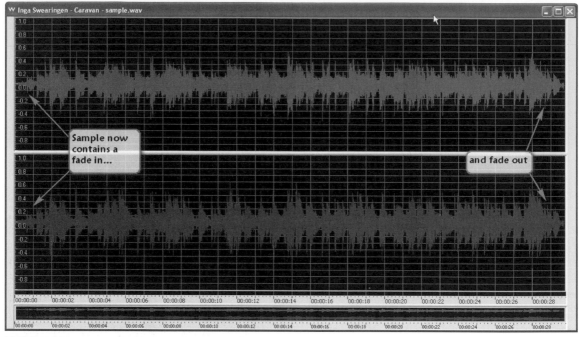

FIGURE 8.7
Song sample with fade in and fade out. (Permission from Goldwave.)

The new file will contain the sample filling up the entire running time of the file—in this case, 30 seconds. At this time, you may want to preview the sample again to verify that it is the best possible representation of the song. If it meets those requirements, it still needs some editing.

To sound like a normal sample, it will need a fade in and fade out. These can be accomplished by highlighting first the beginning section of the song. This will be the area selected for a fade in, from silence to full modulation, so that by the end of the highlighted section, the song is playing at normal volume. The larger the highlighted section, the longer the fade in (see Figure 8.8). Select an appropriate size section for the fade in and use the fade-in tool to reshape the sound. First, highlight the section for the fade in. The appropriate fade-in rate may vary depending on the song and may take some experimentation. Repeat the process in reverse at the end of the sample so that it fades out to silence. This will permanently alter the sample so that no further manipulation is necessary, and the listener will not be required to make any adjustments. Without this editing, the sample would have abrupt entry and exit points and not seem natural.

Then save the file as an MP3 file and upload it to the server. As each web visitor clicks on a link created to the MP3 file, the browser's player will open the file and play it on the visitor's computer. The alternative is to embed the player

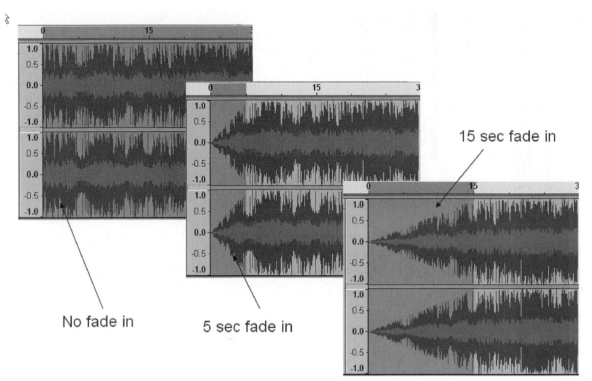

15 sec fade in

No fade in   5 sec fade in

FIGURE 8.8
Illustration of fade-in options.

controls within the web page so that the visitor can click on them to access and
listen to the sample. Most web design software programs include multimedia
controls. This will allow the web designer to place more than one sample on a
page. When setting up the music file on the web page, the options will generally
include the following:

- Do you want this to play automatically when the page opens, or have the
  visitor select play?
- Do you want the song to play once, several times, or loop continuously?
- Do you want embedded user controls?

FIGURE 8.9
Embedded player
user controls.
(Courtesy Nashville
Independent Music,
www.nashvilleim.com.)

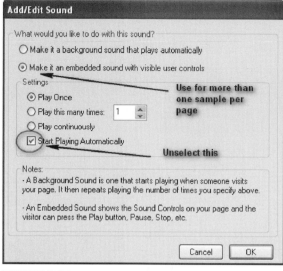

FIGURE 8.10
Example of embedding sound in a web page.

The advantage of having the user's browser open the default media player is that the music will continue to play even if the user moves on to another web page. If the controls are embedded in the page, chances are that the music will quit when the user continues through the web site. However, if there are music files on several pages and all are set to open in the default media player when the page loads, several songs may play at the same time, confusing the visitor.[2]

When including several music samples on one page, it is best not to have any of them play automatically, so that visitors can select if and when they want to listen. A page that automatically plays a sound clip when it's opened may delay the loading process and cause the visitor to wait or give up.

AUDIO EDITING PROGRAMS

Goldwave audio editor, www.goldwave.com

Audacity audio editor, www.audacity.com

Audiobook Cutter, divides longer MP3 files into several smaller files; good for samples, http://audiobookcutter.sourceforge.net

Kristal Audio Engine, www.kreatives.org/kristal

winLAME, converts audio files from one format to another (use in conjunction with an audio editor)

Check www.WM4MB.com for updated lists.

EMBEDDING MUSIC IN FLASH PROGRAMS

Animation, audio, video, or other media that is displayed within a web page is known as embedded media. Embedding media in web pages allows the delivery of an integrated multimedia experience that appears seamless to the user. Although this can be achieved with the above-mentioned streaming process, it is commonly done with a program called Flash by Adobe software. Flash is useful for building small player applications that present audio and video content (Glibertson, 2005). It will allow the user to peruse other content on the page while enjoying the audio file. The flash program could also allow for, say, a slide show, to run while the music is playing.

---

[2] Adding a pop-up player allows the music to play after leaving the page. However, the down side is that, if you click on more than one song, you may have all of them playing at the same time in different players.

MUSIC PLAYERS FOR YOUR WEB SITE

PremiumBeat free MP3 player, www.Premiumbeat.com

The Wimpy Player, www.wimpyplayer.com

Secure-TS Player, fancy skinned secure flash player for under $30, www.tsplayer.com

XSPF Music Player: www.musicplayer.sourceforge.net.

STORING MUSIC OFFSITE

Some musicians bypass the audio process on their web site altogether, opting instead to link to their MySpace page for music sampling. The process of loading audio files to MySpace is simple. First, you must have an artist account, not a personal account. This allows you to feature music. To add up to six songs to your artist profile, go to the manage songs page and begin uploading the files—MySpace only accepts the MP3 format. There are also some music settings that will allow other MySpace users to add one of your songs to their profile page, autoplay the first song when people visit your page, or to randomize the songs played when users visit your page. You can add information about each song, including a graphic and lyrics, in the edit song details section. You can allow users to rate and comment on the song or download it for their personal use. Author Frances Vincent has advised against allowing free downloads—she says you should at least get some marketing information from those who download your music and that is not possible with this system. When this is all set up, you can direct users from your regular web site to your MySpace page just for the music. It is advisable that if the MySpace page is to be used mainly for the music-previewing portion of your web presence, it's best not to load that page up with lots of other competing content, or the visitor may be distracted from the original purpose of previewing the music.

MUSIC VIDEO CLIPS

Music videos made a big splash when they were first popularized by MTV back in 1981. Since then, they have been a staple in the marketing arsenal for any recording artist. Now, YouTube has done for Internet music videos what MTV did for established artists back then. The difference now is that anyone can participate with user-generated content—and it seems like everyone is. YouTube offers a way to share videos in a social networking environment and to create some sense of organization for the thousands of video clips contributed from members. Because video has now become an important marketing tool for musicians, it may be important to include some short video clips in the artist's online presence. You can accomplish this in one of two ways, either by embedding the videos into your artist's own web site or offering links to other video host sites such as YouTube so that the visitor can easily click to watch a video.

Editing Videos

It is unwise to upload raw footage of the artist's live performance without benefit of editing. Most successful video clips are short in length. And a one- or two-minute video will need to feature the best of the best. Three minutes is tops for maintaining interest. Viewer attention span is short, and anything that drones on too long or takes too long to get to "the good stuff" is likely to fail. Viewers will tune it out before giving it a chance. Therefore, it is important to grab people up front. Once the raw footage has been shot and the content of the video is determined, video editing can be done either online, with editors such as eyespot (www.eyespot.com), or with an offline video editor.

Mashups

The latest Web 2.0 video editing hobby is creating *video mashups*. A mashup is the compounding ("mashing") of two or more pieces of complementing web functionalities to create a powerful web application. Mashups originated from the world of pop music where DJs would mix two or more songs together to present a new blended version (www.videomashups.ca). A video mashup is the combination of multiple sources of video—which usually have no relevance with each other—into a derivative work often lampooning its component sources. They are one of the latest genres of mashups and are gaining popularity (Wikipedia). If music videos introduced consumers to fast-paced video cuts and edits, mashups take this art form one step further by allowing the user to develop re-creations of previously released video content. It becomes the perfect new format for developing video projects to promote artists. Some major artists have even used contests to encourage fans to create their own mashups of the artist's materials. This user-generated content then has the potential for spreading virally as proud mashers share their creations with friends.

Often as a pre-release promotion, record labels will distribute raw tracks in free loop bundles for people to download and use to create remixes. The resulting remixed versions of the songs are uploaded to a searchable archive that displays ratings, recommendations and comments from other remixers. More importantly, from a word-of-mouth marketing perspective, this personalized content gets shared through interpersonal networks of friends and family who are online and get remixed versions of an upcoming album release.

http://splinteredchannels.blogs.com/weblog/2005/04/mashup_video.html

- Total Recut is one such social networking site that features video mashups. The site provides tools for creating video mashups, including online video editing, video conversion, and video editing software (www.totalrecut.com/totalrecutdosanddonts.php).
- I Love Music Video.net matches YouTube listings with a user's Last.fm account, making it easy to find videos from their favorite artists (www.ILoveMusicVideo.net).
- Animoto is a service that lets you create a professional-looking music video based on only your uploaded images and music (http://animoto.com).
- Gotuit video mixer gives users the ability to remix videos and produce entirely new video experiences. The Gotuit site claims it's "a whole new way for fans to create their own music videos and share them with friends and family" (www.gotuit.com).
- **Microsoft Popfly** is a mashup development tool that is easy to use for consumer-created mashups but has enough features to attract more serious masher (www.popfly.com).

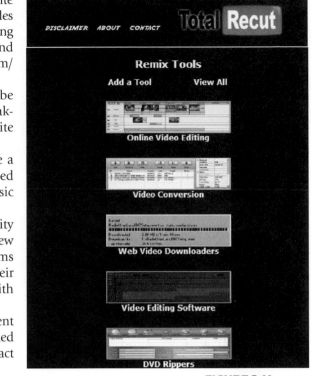

FIGURE 8.11
Video editing and mashup tools from Total Recut. (Courtesy of www.totalrecut.com.)

Placing Videos on Your Web Site

Before deciding to upload videos to the artist site, consider the large amount of server space and bandwidth that videos consume. Bandwidth and file size is a function of the following:

1. Number of frames per second
2. Frame width in pixels
3. Height in pixels

For example, a video shot at 20 frames per second, with a pixel height of 450 and pixel width of 800, will use up 1620 kilobits per second with moderate motion (http://sorenson-usa.com/vbe/index.html).

The next step is to make sure the video is in the correct format for the player that will be used and is one that is compatible with most user systems. Adobe Flash may be the best bet as it works with most browsers and allows for more control over the presentation and the player. If the original file is in some other format, such as Windows Movie Maker, a small conversion program such as Riva FLV Encoder, can be used to convert the video to Flash Video (FLV) (Boutell). Adobe Flash program will allow you to embed the video in the web page with a simple Flash applet. A free alternative, FlowPlayer, is available at the link listed in the accompanying box. Some web design programs also contain features for setting up video embedding.

Once you have the video edited, converted, and the appropriate audio track coupled with it, upload the file to the server. It may be wise to create a new folder on the server to hold video file and the *applet* (a small Java program) that runs the video. Then set up the player's attributes: size, look, controls, skins, and so on. It might be a good idea to feature a link on your web site for visitors to download a copy of the latest flash player.

VIDEO EDITING TOOLS

Riva FLV encoder (free trial), www.rivavx.com/?encoder

ConvertTube, www.converttube.com

Flvix, a simple free online video converter, www.flvix.com

Gaffitti, online video editor, http://graffiti.vidavee.com

FlowPlayer, http://sourceforge.net/projects/flowplayer

Tutorial on how to add video to a web site,
   www.boutell.com/newfaq/creating/video.html

Tutorial on how to add video to a web site,
   www.2createawebsite.com/enhance/adding-video.html

Flash player, www.macromedia.com/software/flash/about

Updated information available at www.WM4MB.com.

FIGURE 8.12
Embed a video player
in your web page.

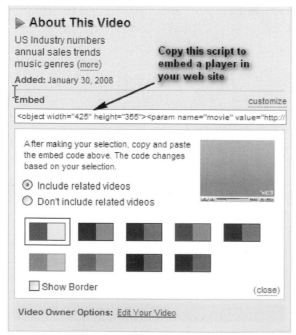

Pointing Visitors to a Video Hosting Service

An easy alternative to hosting your own videos is to use one of the video hosting services such as YouTube to store the video. YouTube offers two ways to promote a YouTube video on your web site: a link to the page with the video or HTML text that embeds the YouTube video player in your web page. Each video uploaded to your YouTube account comes with a piece of HTML code written so that you can then place in your web page to create an onsite player.

Google Video, StreamLoad, and imeem are three of the many other video hosting services. Imeem allows you to upload videos to your profile and share them with others in a variety of social network situations. For video files, imeem supports most formats but recommends .MPEG, .MOV, .FLV, and .AVI for best results. The suggested setting for a video file is 400 × 300 with mpeg-4 video and mp3 audio. Just as with YouTube, imeem also allows users to create video channels. Chapter 12 has a section on promoting your videos on YouTube.

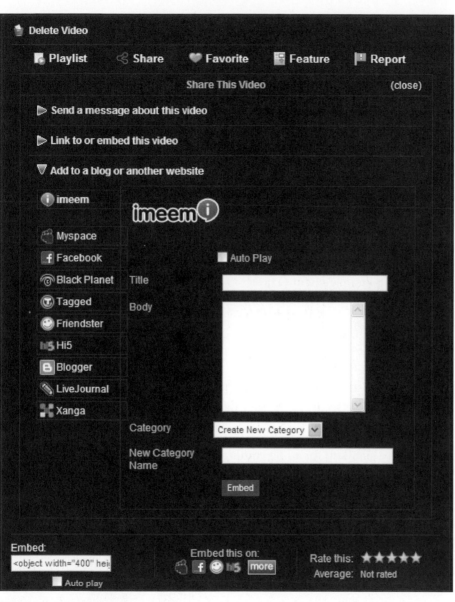

FIGURE 8.13
The imeem video page
with widgets to add the
video to other social
networking sites.

CONCLUSION

Today's Web 2.0 atmosphere offers many opportunities to display and share audio and video. Online consumers have come to expect a multimedia experience, and every music and entertainment web site should offer them one. The wise music marketer will take advantage of all options, including adding audio and video to the primary web site and creating pages on social networking sites that include audio and video of the artist. Music samples should be provided in easy-to-find and well-traveled places. Mashups are becoming popular now and can showcase musical talent in new and exciting ways. With today's ADD (attention deficit disorder) generation (Nalts, 2007), videos need to be short, and the mashup is a perfect format for that.

GLOSSARY

Advanced Audio Coding (AAC) – A standardized encoding scheme for digital audio, promoted as the successor to the MP3 format. AAC generally achieves better sound quality than MP3 at the same bitrate. Used by Apple's entertainment products and mostly with DRM in place (although that is changing).

Applet – A special type of Java program that can be included in an HTML page. Web browsers, which are usually equipped with Java virtual machines, can run the applets to perform interactive graphics, games, calculators, and so on.

Audio Home Recording Act of 1992 – Established a number of important precedents in U.S. copyright law that defined the debate between audio and video device makers and the content industries, requiring all digital audio recording devices sold, manufactured, or imported in the United States (excluding professional audio equipment) to include the Serial Copy Management System.

Digital rights management (DRM) – A systematic approach to copyright protection for digital media whose purpose is to prevent illegal distribution of paid content over the Internet (Bitpipe.com).

Gracenote – Primarily a music recognition service that works with your computer's software player application to identify an audio CD and display the artist, title, tracklist, and other information.

Mashup – A mixture of content or elements; the compounding ("mashing") of two or more pieces of complementing web functionalities (audio or video) to create a powerful web application.

MP3 file – Stands for MPEG (Motion Pictures Expert Group) Audio Layer III, and it is a standard for audio compression that makes any music file smaller with little loss of sound quality.

Serial Copyright Management System – Created in response to the digital audiotape (DAT) invention, in order to prevent DAT recorders from making second-generation or serial copies. SCMS sets a "copy" bit in all copies, which prevents anyone from making further copies of those first copies. It does not, however, limit the number of first-generation copies made from a master. SCMS was an early form of digital rights management (DRM). It was also included in consumer MiniDisc and DCC players and recorders.

Streaming audio – "A one-way audio transmission over a data network. It is widely used on the web as well as private intranets to deliver audio on demand or an audio broadcast (Internet radio). Unlike sound files (WAV, MP3, etc.) that are played after they are downloaded, streaming audio is played within a few seconds of requesting it, and the data is not stored permanently in the computer" (ZDNet).

Video mashup – The combination of multiple sources of video—which usually have no relevance with each other—into a derivative work often lampooning its component sources.

Wave file – Short for Waveform audio format, a Microsoft and IBM audio file format standard for storing audio on PCs.

Windows Media Audio – An audio data compression technology developed by Microsoft.

REFERENCES AND FURTHER READING

Ambliozone.

Amib. How to choose the appropriate lossy audio compression format. Everything2.com, http://everything2.com/index.pl?node_id=1683222.

Cornell University Information Technology, http://atc.cit.cornell.edu/course/streaming/index.cfm.

Gilbertson, Scott. (2005, November 9). Building an embedded MP3 player in Flash. WebMonkey, http://www.webmonkey.com/webmonkey/05/46/index2a.html?tw=multimedia.

Haring, John. (2008). (Personal interview).

Hinchcliffe, Dion. (2007). More results on use of Web 2.0 in business emerge. ZNet, http://blogs.zdnet.com/Hinchcliffe/?p=103 (April 3).

http://www.webmonkey.com/webmonkey/05/46/index2a_page5.html?tw=multimedia.

Nalts. (2007). How to promote your videos on YouTube. http://nalts.wordpress.com/2007/06/01/how-to-promote-your-videos-on-youtube (June 1).

Nevue, David. (2005). *How to promote your music successfully on the Internet,* The Music Business Academy, www.musicbusinessacademy.com.

PC Magazine online. www.pcmag.com/encyclopedia_term, www.pcmag.com/encyclopedia_term/0,2542,t=streaming+audio&i=52132,00.asp.

Potts, Daniel. (2002). Audio compression formats compared. Australian PC World. http://www.pcworld.idg.com.au/index.php/id;1090733979;fp;2;fpid;206 (July 31).

Recording Industry Association of America v. Diamond Multimedia Systems, Inc., http://cyber.law.harvard.edu/property00/MP3/rio.html.

www.videomashups.ca.

Vincent, Frances. (2007). *MySpace for Musicians.* Thomson Course Technology, Boston, MA.

CHAPTER 9

E-commerce: Product Ordering and Fulfillment

For most artists and record labels, the ultimate goal is selling a product, especially recordings and concert tickets. To do this, it is probably necessary to get into e-commerce: having a retail presence on the Web. The first decision involves whether to "set up shop" or leave it to the experts.

There are many online retail stores that handle either physical product, digital downloads, or both. Should you decide to do it yourself, there are several online services that handle the complex portions of self-distribution, including financial transactions, setting up the web site storefront, and inventory management and handling. Some of these services simply offer software to interface with your artist's web site and use a database to manage shipping information, or they simply deal in financial transactions.

DOING IT YOURSELF

Doing it yourself requires several components in the process of engaging in commercial transactions: processing orders online, providing financial security, handling the financial transaction, inventory management, and shipping out orders.

Fulfillment

Fulfillment is defined as order processing that includes documenting when an order was received, when and how it was shipped, and when and how it was paid for. Record labels, artists, and their managers need to weigh the options when deciding whether to handle their own fulfillment. It requires persistent attention to the web site and prompt follow up on all orders received. If an artist is on the road touring, fulfillment should be left to a third party to handle. Advances in computer technology have streamlined the fulfillment process.

135

Table 9.1	Comparison of Options for E-commerce

Comparison of Using Fulfillment Services versus DIY

	Advantages	Disadvantages
Doing It Yourself	Keep more of the profit per unit No up-front costs or monthly fees Good for those with more time than money	Requires constant commitment Requires knowledge of web development and security systems Must handle storage and shipping Customers may be reluctant to provide credit card numbers
Using shopping cart software or services	Reliable and secure transactions Inventory management is more organized Suitable for artists with limited Internet skills but who are available to handle fulfillment	Up-front costs to set up shopping cart Monthly or transaction fees cut in to profit Still must handle storage and shipping
Using full-service fulfillment	They are responsible for fulfillment, order processing, inventory management Suitable for artists who are busy or on the road Can provide marketing services	Take a much larger percentage of sales, much like a retail store

With the right software program in place, much of the order processing can be automated, from keeping tabs on inventory levels to actually printing out shipping labels and bar codes for the delivery services.

Financial Transactions

CREDIT CARDS

To set up shop, it is necessary to process transactions and collect money from your customers. Credit cards are the most popular form of financial exchange online because of their convenience and speedy processing. Years ago, mail-order businesses mostly requested money orders or cashiers checks; some would take personal checks but would then wait until the check had cleared before sending out the product. Back then it was common to see the disclaimer "allow four to six weeks for delivery." In today's immediate gratification society, four to six weeks is not an acceptable time frame for most customers. Credit cards increase impulse buys. In *"How to Promote Your Music Successfully on the Internet,"* David Nevue (2007) stated, "To run a successful business on the Internet, credit card acceptance is an absolute must." However, credit card processing is not without costs and requires an elaborate setup. There are web service companies now that provide the credit card processing services along with the software to integrate into your web site. To set up a site to accept credit cards, your company must have a merchant bank account, security and encryption measures in place (*secure socket layer* server or SSL), credit card verification services (also called a payment gateway), a shopping cart page, and the software to process and track orders and shipments. In addition, some customers are apprehensive about giving out their credit card numbers to an unknown vendor on a small web site and prefer to use a more reputable retailer or more secure service such as PayPal.

PAYPAL

PayPal is an online service that allows registered users to transfer funds to and from bank accounts set up as their PayPal accounts. It also allows nonregistered users to make a payment to a registered user via a major credit card. The payee will feel more secure providing his or her credit card number to PayPal than the small, unknown vendor who is selling items on the Internet. PayPal then credits the payment to the vendor's account. The vendor pays a service fee of 30 cents plus a small percentage of each transaction. The service also offers online shopping cart services.

GOOGLE CHECKOUT

New on the scene is Google Checkout, a service offered by Google. When combined with Google's AdWords program (see Chapter 11), the costs are greatly reduced, with merchants waiving the monthly transaction fees on $10 in sales value for every dollar spent on advertising with Google.

Keep in mind that doing it yourself involves a commitment to maintain accurate accounts of inventory on the web site and to promptly respond to each transaction with a confirmation e-mail and shipping information. It also requires being ready to pack up and ship products out the door on a frequent and consistent basis. For musicians who are on the road, this responsibility is best left to the experts.

For handling e-commerce on the web site, it is necessary to follow these recommendations:

1. Provide thorough product descriptions, including graphics.
2. Prominently display the product name and price. If several formats are available, clearly identify the format.
3. Make it easy for the customer to purchase; the fewer clicks, the better. Make the "buy" button obvious.
4. Make sure the customer knows when the order is completed.
5. Once the order is placed, send an immediate e-mail confirmation.
6. Make sure the orders go out as quickly as possible.

7. Make sure the customer knows how long it will take to receive the order.

8. Make sure you have the inventory to fill the orders. If you are out of stock, modify the storefront page immediately to reflect this fact.

Shopping Carts

The *online shopping cart* offers customers a convenient and straightforward way to purchase items from a web vendor. Webopedia describes the shopping cart as

> a piece of software [or service] that acts as an online store's catalog and ordering process. Typically, a shopping cart is the interface between a company's Web site and its deeper infrastructure, allowing consumers to select merchandise; review what they have selected; make necessary modifications or additions; and purchase the merchandise.

The shopping cart software also offers the vendor seamless management of inventory, revenue and tax collection, and shipping information. Many shopping cart software vendors and services have the capability to design the shopping cart page so that it is integrated into the look and feel of the rest of the web site.

There are two options for integrating a shopping cart into the artist's web site: (1) software programs that allow the web designer to specific parameters and create the cart or (2) online services that offer inventory management features and the ability to customize their shopping cart to conform to the design of your web site. In the article "E-commerce Shopping Cart Review," author Dan Wellman evaluates several software programs, including Nopcart, Shop Script Free, The Zen Shopping Cart, and Comersus Cart Life. The advantage of using a software program is that there are no commission fees to pay, only the initial cost of the software. This arrangement may be preferable for companies that do a great deal of business and can dedicate an employee or two to monitor and maintain the interface.

FIGURE 9.1

The second option, hiring an online service to handle and host the shopping cart services, can reduce the possibility of complications as most services have security and transaction management procedures covered. These services also provide database management and keep records of transactions for the vendor.

The web site EarlyImpact.com outlines the features one must examine when selecting which shopping cart service to adopt. The first is to focus on features that are important to your needs and eliminate programs and services that don't match those needs. Second, visit live stores that use each of the services or programs you are considering. Walk through the customer experience by shopping and adding things to the shopping cart. Continue exploring by manipulating the items in the cart. Is it easy to remove, add, change quantities, and so on? The third aspect of comparing options is to determine what the total cost will be for each of the possibilities. Some offer a lower setup fee, or monthly rate, but charge more commission per transaction. This may be suitable for low-volume business. But at higher sales volumes, it may be more cost effective to pay a higher up-front setup fee or higher monthly fees and make up the savings with lower commission rates.

Next, review store management tools, which can be done through product demonstrations. Note how easy or complicated it will be to perform inventory management and shipping functions such as processing orders, authorizing funds, sending out automated e-mails to the customer, manually changing orders, batch processing multiple orders, and so forth. The article also suggests looking up published reviews of the product and evaluating product support.

TOP 12 PROBLEMS WITH SHOPPING CART DESIGN

In 2002, Barbara Chaparro conducted research on various shopping cart programs and published the *Top Ten Mistakes of Shopping Cart Design*. The study was repeated in 2007 to determine if any of the problems had been addressed. The following summarizes the findings:

1. Calling it something other than a shopping cart. Whereas the earlier study found that U.S. consumers are accustomed to the term *shopping cart*, the later study revealed that the term was used more literally in some other countries to mean only the physical shopping carts in bricks-and-mortar stores.
2. Requiring users to click the BUY button in order to add something to their cart. As in physical stores, the shopping cart should be a place to put items

Table 9.2 **Examples of Some Shopping Cart/Transactional Services**

	1ShoppingCart	Yahoo! Small Business	PayPal	BizHosting
Setup	$0	$50	$0	$14.95
Monthly	$29–$79	$39 and up	$0	$21.95 and
	$0	1.5%	2.9% +	up
Transactional			$0.30	$0
Digital downloads	Yes, but limited	Yes	N/A	Yes
Payment methods supported	All major credit cards, PayPal, eChecks	All major credit cards, PayPal	All major credit cards, debit cards, direct account transfer	All major credit cards, PayPal

that may or may not be purchased upon checkout, without requiring customers to commit to each item at the time they place the item in the cart. An "add to cart" button is more appropriate. Some programs now offer a "wish list" function for customers who prefer to purchase at a later time.

3. Giving little to no visual feedback that an item has actually been added to the cart. With some more ambiguous programs, the customer may not be sure their item has been place in the cart, only to discover later that they have three or four of the same object added to the shopping cart list when they prepare to check out. The author states that as of 2007, about two-thirds of sites take the user to a shopping cart page when an item is added.

4. Forcing the shopper to view the cart every time an item is added to the cart. This takes the shopper away from shopping mode and may decrease overall sales. If the cart information is included on the shopping page, such as with Amazon.com, it is not necessary to take the customer to the cart page after each item is added.

5. Asking the user to buy other related items before adding an item to the cart. Again, disrupting the flow of shopping at this point in the shopping experience is not a good idea. It is better to show related products either just after an item is placed in the cart, or better yet, just before the customer checks out.

6. Requiring users to register before adding an item to the cart. Customers report that they do not like to provide personal information until they are ready to check out. In addition to the intrusion, it also disrupts the shopping flow.

7. Requiring a user to change the quantity to zero to remove something from the cart. It is much better to provide a remove or delete button next to each item in the shopping cart, yet the "change quantity to zero" function still persists on 15% of sites tested.

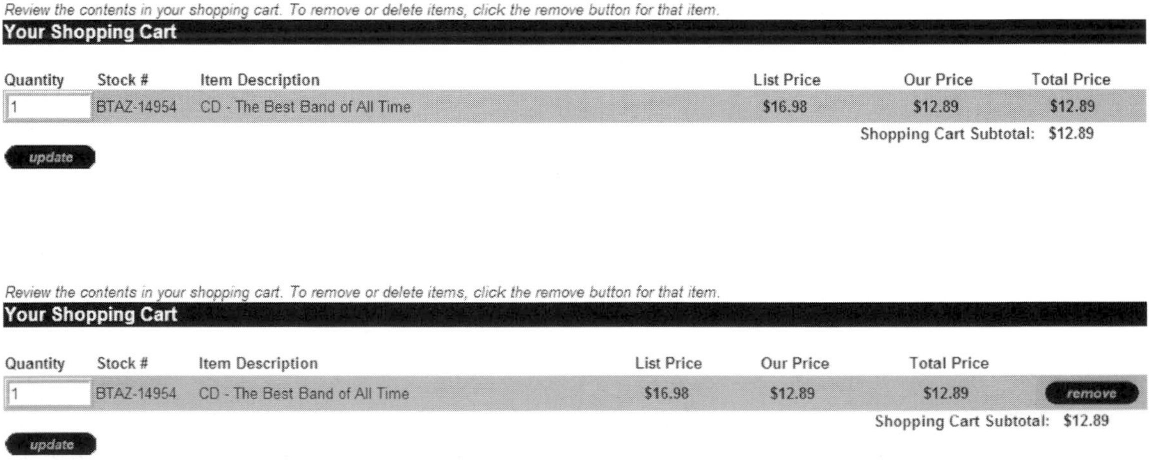

Review the contents in your shopping cart. To remove or delete items, click the remove button for that item.

Your Shopping Cart

Quantity	Stock #	Item Description	List Price	Our Price	Total Price
1	BTAZ-14954	CD - The Best Band of All Time	$16.98	$12.89	$12.89

Shopping Cart Subtotal: $12.89

update

Review the contents in your shopping cart. To remove or delete items, click the remove button for that item.

Your Shopping Cart

Quantity	Stock #	Item Description	List Price	Our Price	Total Price	
1	BTAZ-14954	CD - The Best Band of All Time	$16.98	$12.89	$12.89	*remove*

Shopping Cart Subtotal: $12.89

update

FIGURE 9.2
Portion of shopping cart shown without and with the remove button.

8. Including written instructions to update the items in the cart. Today's online shoppers do not bother reading instructions. Using the shopping cart should be intuitive enough that they can focus on shopping instead of managing the cart.

9. Requiring a user to scroll down to find important buttons on the cart page. The update cart button, as well as the checkout button should probably be located at the top *and* the bottom of the shopping cart page so that customers with a lengthy list of products will not have to scroll to find them.

10. Requiring a user to enter shipping, billing, and personal information before knowing the final costs, including shipping and tax. Some of the newer sites open a separate window to compute shipping and tax, dependent on the customer's shipping address. But nearly half of the sites tested still required most of this information before showing the final total.

The 2007 study included some new issues not covered in the original study.

11. Security is an issue with consumers, so verified secured sites are necessary.

12. Out-of-stock items were a problem in that some consumers were not notified that their items were out of stock until they reached the check-out page.

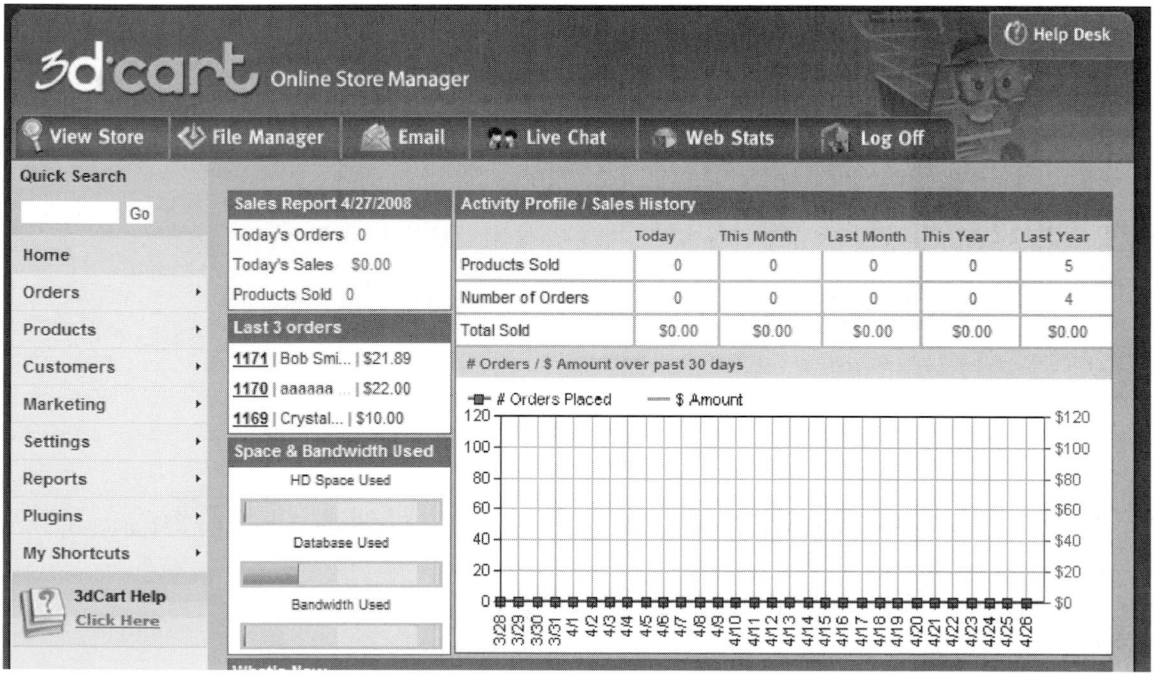

FIGURE 9.3
Example of what information the vendor has access to. (Courtesy of http://www.3dcart.com, by permission.)

ONLINE RETAILERS

There are advantages to signing on with an online retailer. The major decision is whether you want to handle the money and product fulfillment yourself, or whether it is worth sharing some of the profit with an online retailer and letting the retailer manage transactions and shipment. Online retailers can provide benefits such as marketing, reputation, and traffic—much like a mall offers shopping traffic to its tenants.

Online Storefronts

There are many online retailers that offer services to small vendors (musicians and indie labels), including product fulfillment, collection of revenue, and some online promotions. The most popular online retailers for music are iTunes, CDBaby, Amazon.com, AmieStreet, and SNOCAP. In addition to the retailers, there are *aggregators*, digital music wholesalers who aggregate music from many sources and distribute to a multitude of retailers.

- *CDBaby*. CDBaby is quite popular with independent and developing artists. David Nevue (2007) stated, "CD Baby is the largest seller of independent music." The setup cost per title is $35 to cover setting up the account and web page. CDBaby's web pages include song samples, artwork, artist bio, customer reviews, and a link to the artist's web site. CDBaby does not charge a monthly fee, only the $35 setup fee and $4 per unit sold. They ask for five CDs from unproven artists to start the process. As those are sold, the artist is notified and more CDs are requested. CDBaby also features artists, provides bar codes, and reports sales data to SoundScan.
- *Amazon.com*. Amazon is a bit more expensive but is very good for raising your profile. Music is submitted for review, which means there is some quality control. Labels should set up an Advantage account, which provides an online storefront for the label's catalog. The advantage of Amazon is the extremely high traffic generated by the site. The disadvantages include the fact that Amazon takes a large portion of the profit from each sale and charges an annual fee to participate. Amazon has also launched a digital download service, available to independent musicians.
- *iTunes*. The number one retailer also has the lion's share of the digital download market. While iTunes is set up for direct accounts with established labels, an independent artist or small indie label must go through a distributor such as CDBaby, Tunecore, or Songcast to get songs placed on iTunes (Shambro). iTunes does not like to receive the same content from more than one distributor, so if you are working through several distributors or aggregators, designate one as your representative to iTunes.
- *SNOCAP*. Founded by Shawn Fanning of Napster fame, SNOCAP teamed up with MySpace to sell music from unsigned artists through the MySpace web site. The struggling company was then acquired by imeem in February 2008. At the time of this writing, an artist limited account is free and SNOCAP charges 39 cents per download. It is first necessary to set up an account with SNOCAP and upload your music. Then you can click on

"post to web" to get HTML language to create an imbedded SNOCAP store in MySpace, ReverbNation, or Shoutlife, with more coming soon. In addition, you can make your SNOCAP store display on any web site that accepts HTML code, which would include the artist's primary web site. SNOCAP's web site states:

1. Log into your SNOCAP Artist Account.
2. Click on the "SNOCAP MyStore" tab.
3. Click on the "Post to Web" tab located below your store name.
4. You will see the HTML code in a text box. Copy and paste this code in your personal web site (or your Blog, MySpace Comments, Message Board, e-mails, etc.)

```
<embed src="http://void.snocap.com/s/T3-31324-4FWCP32K8L-Z/" width="425"
height="300" wmode="transparent" style="background: url(http://void.snocap.
com/b/T3-31324-4FWCP32K8L-Z/);"/>
```

Snocap is also supported on ReverbNation and ShoutLife.

- *CNET's Download.com.* The music area of download.com offers artists the chance to upload their music to the free music database (http://music. download.com/3750-21_32-40.html). Even major labels are participating in offering free samples, downloads, and streaming. Click on "submit software" at the bottom of the main page to get to www.upload.com.

- *AmieStreet.* AmieStreet combines digital music retailing with social networking to provide an atmosphere of exploration and discovery of new music. The price of music increases as its popularity rises, with fans participating in the profit sharing. This provides an incentive for users to spread the word about new recordings and artists. Artists collect 70% of the monies collected from each sale after the song has made $5 (http://www. amiestreet.com/). On the home page, click on "sell your music" in the menu bar underneath the masthead.

- *The Orchard.* The Orchard claims to be the world's largest distributor of digital music. They are one of the most prominent *digital aggregators*, or middlemen between record labels and online retailers. The company distributes to iTunes, emusic, Napster, Rhapsody, MusicMatch and MusicNet. It also supplies music to Verizon's wireless service V Cast. The Orchard is a bit more difficult to get involved with on a small scale and tends to be more practical for record labels with a substantial catalog than for individual artists.

- Other *aggregators* include Tunecore (http://www.tunecore.com/), SongCast (http://www.songcastmusic.com/), and IRIS (www.irisdistribution.com). TuneCore charges $0.99 per track to register, $0.99 per store per album, and $19.98 per album per year storage and maintenance. The service is open to anyone. SongCast charges about $6 per month for an account that allows for unlimited tracks, and a $25 set up fee per album. Distribution is provided to iTunes, Rhapsody, Amazon, Emusic and Napster. IRIS focuses on digital distribution for independent artist and indie labels.

OTHER ARTIST-RELATED PRODUCTS

In addition to selling recorded music, artists often sell branded T-shirts, hats, coffee mugs, bumper stickers, and other items commonly referred to as *swag* (stuff we all get). Although these items are often sold at live performances, they can also be offered for sale online. For artists who do not want to personally handle the manufacturing and fulfillment of these items, there are companies available online who will handle swag e-commerce. Café Press is one option for independent artists who don't want to invest in a large inventory of products but who want to provide options for fans. There are no startup costs for a basic account, and the company prints on demand with no minimum order, but the base price is high, leaving little room for markup. Another online site, Zazzle, offers a feature for setting up your own shop and posting it on MySpace and Facebook. Zazzle also prints to order and specializes in band merchandise. For larger volume merchandising, PrintMojo charges a much smaller fee per item but does require a minimum order of 25 shirts (www.printmojo.com).

FIGURE 9.4
Set up your own online merchandising shop at www.zazzle.com. (Courtesy Zazzle.)

Jump Drives

New to the scene is the use of jump drives or thumb drives to store and sell music. This reusable media allows artists to sell extended musical works in the MP3 format loaded on a jump drive, along with videos, graphics, and other artist-related content. We are beginning to see widgets designed to automatically load the music content into the consumer's music library for transfer to other portable devices. Several companies offer bulk quantities of jump drives complete with the artist's logo for under $6 each. These are ideal for selling music in situations where physical product is still advantageous (such as live shows) because the reusable jump drives add value to the product and are better for the environment than optical discs. To help stimulate sales at live shows, it is recommended that the drives be sold with a lanyard, so that the consumer can wear them around their neck at the show, like a backstage pass. The visibility of these may entice other attendees to purchase their own. Otherwise, they would end up in the consumers' pockets and lose the viral impact.

THE LAST WORD ON E-COMMERCE

While online stores are starting to become more consumer-friendly, the online shopping process is still in the evolutionary stage of development—unlike traditional retail stores that have been researched for decades. According to FutureNow's "2007 Retail Customer Experience Study," mystery shoppers were sent to more than 300 retail web sites and reported the following statistics for December 2007 (From eMarketer.com):

- 74% offered estimated delivery times
- 61% did not offer any information on the product page regarding in-stock availability
- (Only) 58% correctly answered an e-mail question within 24 hours
- 52% of retailers had physical stores; only 10% of all retailers offered in-store pickup of orders
- 43% offered free shipping
- 42% provided shipping costs early in the checkout process
- 35% had a checkout process with more than four steps
- 33% offered customer reviews

GLOSSARY

Aggregator – A distribution company that acts as a middleman between record labels and online retailers. Examples include (Independent Online Distribution Alliance) IODA, The Orchard, and IRIS Distribution.

Fulfillment – Processes necessary to receive, service, and track orders sold via direct marketing. "The primary functions of fulfillment systems are (1) to respond quickly and correctly to an order by delivering the item ordered, (2) to maintain customer records, (3) to send invoices and to record payments, (4) to respond to customer inquiries and complaints and resolve problems, and (5) to produce purchase and payment information on an individual customer basis." (Answers.com)

PayPal – PayPal is an online service that allows registered users to transfer funds to and from bank accounts and credit cards that they set up for their PayPal accounts. It also allows nonregistered users to make a payment to a registered user via a major credit card.

Product fulfillment – The gathering of orders from a sales transaction and the process of completing the order through delivery of the ordered merchandise.

Secure socket layer (SSL) – A protocol developed by Netscape for encrypting private documents for secure transmission via the Internet.

Shopping cart – A piece of software that acts as an online store's catalog and ordering process, allowing consumers to select merchandise, review what they have selected, make necessary modifications or additions, and purchase the merchandise.

Swag – Souvenir or promotional items associated with a product or brand, or short for "stuff we all get."

REFERENCES AND FURTHER READING

Answers.com.

CafePress.com.

Chaparro, Barbara. (2002). Top ten mistakes of shopping cart design, http://psychology.wichita.edu/surl/usabilitynews/42/shoppingcart.htm(July).

eMarketer.com. (2007). Many web retailers miss the basics, www.emarketer.com/Article.aspx?id=1005666 (December 3).

Naidu, Shivashankar, and Chaparro, Barbara. (2007). Top ten mistakes of shopping cart design revisited: a survey of 500 top e-commerce web sites, http://psychology.wichita.edu/surl/usabilitynews/92/shoppingcart.html.

Nevue, David. (2007). How to promote your music successfully on the Internet. The Music Business Academy, www.musicbizacademy.com.

Review and compare shopping carts: how to choose the best ecommerce software. Navigating software reviews and making a good decision, www.earlyimpact.com.

Shambro, Joe. Getting sold on iTunes. About.com. http://homerecording.about.com/od/duplicatingdistributing/a/Get_On_iTunes.htm.

Wellman, Dan. (2005). E-commerce shopping cart review. Web Hosting News. http://webhosting.devshed.com/c/a/Web-Hosting-News/Ecommerce-Shopping-Cart-Review/

CHAPTER 10
Finding Your Online Market

Perhaps the most important aspect of marketing involves finding and getting to know your market. On the most basic level, markets can be segmented into three sections: (1) fans and current users, (2) potential fans and users, and (3) those people who are not considered part of the target market. Perhaps this third group includes people who cannot or will not consume your products. For music, that may mean people who do not particularly care for the genre that your artist represents. It may include people who do not consume music, people who are unwilling to pay, and those without access to become consumers. For example, there have been mid-level and star-level recording artists who have enjoyed a fan base that includes very young children, for whom regular concert attendance would be impractical. Perhaps the concert is too late, too loud and rowdy, or is restricted to those of legal drinking age. Artists who are aware that they have a following of small children have been known to add a special matinee performance to the schedule to accommodate families with small children. This is an example of why it is important to learn as much as possible about your market so that you can adjust your marketing strategy to increase the likelihood of success. This chapter focuses on the first two market sections: current fans and potential fans.

MARKET SEGMENTATION

The basic goal of *market segmentation* (subdividing a market) is to determine the target market. Because some markets are so complex and composed of people with different needs and preferences, markets are typically subdivided so that promotional efforts can be customized—tailored to fit the particular submarket or segment. For most products, the total potential market is too diverse or heterogeneous to be treated as a single market. To solve this problem, markets are divided into submarkets called *market segments*. Market segmentation is defined as the process of dividing a large market into smaller segments of consumers who have similar characteristics, behaviors, wants, or needs. The resulting segments are homogenous with respect to characteristics that are most vital to the marketing efforts. That means that members of the segment have enough in common with each other that customized messages can be more effective. This segmentation may be made based on gender, age group, purchase occasion, or benefits sought. Or they may be segmented strictly according to their needs or preferences for particular products. The Internet has revolutionized the way markets are segmented because so much more data are available on consumers' interests and purchase behavior.

To be successful, segmentation must meet these criteria:

1. *Substantiality.* The segments must be large enough to justify the costs of marketing to that particular segment. Because the Internet allows for more effective targeting of markets and the cost of reaching each member is lower, this is starting to change so that smaller and smaller "segments" are worth individual attention.

2. *Measurability.* Marketers must be able to analyze the segment and develop an understanding of their characteristics. The result is that marketing decisions are made based on knowledge gained from analyzing the segment. The Internet has allowed for data mining and *behavioral targeting,* creating market segments based on what Internet users purchase online and what types of sites they visit. For example, based on weather reports and restaurant listings online, a search engine company can determine where someone lives. And based on searches they have conducted on the Web and what keywords they have used, the company can determine what products that person might be interested in receiving information about (Jesdanun, 2007).

3. *Accessibility.* The segment must be reachable through existing channels of communication and distribution. With web marketing, accessibility is no longer the problem it once was. The only limitations are with the technology—some users still have dial-up Internet service and care should be taken not to overload them with slow-loading files.

4. *Responsiveness.* The segment must have the potential to respond to the marketing efforts in a positive way, by purchasing the product. Internet users are becoming more comfortable with purchasing products online. Services such as PayPal have increased the willingness of consumers to deal with small, unknown web vendors with confidence and ease. However, there are still some segments of the market whose members are reluctant to order products online for security reasons, because they do not possess a credit card, or because they are unwilling to provide the credit card number to the vendor.

Market Segments

The process of segmenting markets is done in stages. In the first step, segmentation variables are selected and the market is separated along those partitions. The most appropriate variables for segmentation will vary from product to product. The appropriateness of each segmentation factor is determined by its relevance to the situation. For example, age may be a significant factor for products that are age related, such as acne medicine or denture adhesive. For other products, age may not be as significant. After the salient market characteristics are determined and the market is segmented, each segment is then profiled to determine its distinctive demographic and behavioral characteristics. Then the segment is analyzed to determine its potential for sales. The company's target markets are chosen from among the segments determined at this stage.

There is no single correct way to segment markets. Segmentation must be done in a way that maximizes marketing potential. This is done by successfully targeting each market segment with a uniquely tailored plan—one that addresses the particular needs of the segment. Markets are most commonly segmented based on a combination of geographics, demographics, personality or psychographics, and actual purchase behavior. Traditional marketing has relied on these types of segments or combinations of them (using some demographics combined with psychographics), but it has been evolving to include more purchase behavior as technology provides a means for measurement. Behavior segmentation is a more effective way to segment markets, because it is more closely aligned with propensity to consume the product of interest.

We've Got Our Segments, Now What?

Target marketing involves identifying a market segment to "go after" with some sort of marketing campaign. The decisions about what that campaign should involve are made based on information gathered about the market segment. Table 10.1 lists some of the information you might want to know about your target market.

Table 10.1	Application of Marketing Research to Strategy	
	What We Know about Our Target Market	**Marketing Factor Based on That Knowledge**
1	Age, gender, ethnicity, hometown. To know more about the demographic and geographic makeup of our market.	Where can we find our market? What categories do they fall in to?
2	How did you hear about this CD? What influenced you to buy this CD? (e.g., radio, TV, in-store, concert, magazine, etc.)	Which of our marketing efforts are most successful in reaching our customers and which were wasted money?
3	How many other albums do you own by this same artist?	Loyalty: How loyal are you to this artist?

(Continued)

Table 10.1	Application of Marketing Research to Strategy—cont'd	
	What We Know about Our Target Market	**Marketing Factor Based on That Knowledge**
4	What do you read? Watch? Listen to?	How can we best reach you with future advertising and media placement?
5	How much do you buy online? Mail order? Catalog? Toll-free phone numbers?	To what extent should we be using alternative distribution methods?
6	What is your favorite beverage? Sneaker company? Car or truck? Music store?	What companies should we team up with for promotional tie-ins?
7	How much time do you spend online? Shopping? Out to dinner? At the mall?	Where can we find you to target you with our marketing efforts?
8	Do you own a DVD or video game player, computer, cell phone? Do you go to the movies, clubs?	What is competing with music for your entertainment dollar?
9	Do you own or use any of the following consumer electronic devices: DVD, MP3 player, CD-R burner, computer music software, multiuse cell phone, etc.?	What configurations should we use to deliver our recorded music and marketing messages?
10	What are your hobbies and interests?	Where else can we find you when you're not listening to our music or doing any of the above activities?

Once you know these characteristics about your target market, you can decide what to say and where to say it.

Marketing Research Companies

There are several marketing research firms that collect and provide data and analyses on consumer groups. Many of these specialize in Internet consumers, and several also cover the recording industry and technology. Generally, the reports are for sale, and it may be worth purchasing one or two before starting out on a new venture for an artist or a label. These companies are also contracted by the various industry associations to conduct specialized research that is then made available to association members.

Forrester Research is one of the major market research firms focused on the Internet and technology; the company conducts research for the recording industry on all aspects of music and the Internet. Forrester also offers custom research and consulting services to its clients. *Jupitermedia* Corporation is a top provider of original information, images, research, and events for information technology, business, and creative professionals. The associations often hire Jupitermedia to conduct and report on online music consumers. *Edison Media Research* is a leader in political, radio, and music industry research with clients that include major labels and broadcast groups. *Music Forecasting* does custom research projects on artist imaging and positioning.

The *NPD Group*, recently acquired *by Ipsos,* provides marketing research services through a combination of point-of-sales data and information derived from a consumer panel. NPD's research covers music, movies, software, technologies, video games, and many other product groups. *ComScore* offers consulting and research services to clients in the entertainment and technology industries and conducts audience measurements on web site usage through its *Media Metrix* division. Taylor Nelson Sofres, a UK firm, provides both syndicated and custom research of media usage and consumer behavior. Based in France, IPSOS is a global group of researchers providing survey-based research on consumer behavior. *BigChampagne* (owned by Clear Channel) tracks online P2P usage and reports, among other things, the most popular songs on P2P networks. *ResearchMusic* is an online music service that also offers free music market research to registered customers. OTX is an online consumer research and consulting firm.

For customized research, the company Indie Marketer provides market analysis services and helps direct marketing campaigns to advance artist careers (www. indiemarketer.com).

TRACKING CONSUMER BEHAVIOR ON THE WEB

Before the emergence of the Internet, marketing researchers used a variety of techniques to learn more about consumer behavior. Many of these studies were not comprehensive, meaning that shopping behavior may be measured on one group of consumers while advertising exposure is measured on another. It was difficult to conduct a comprehensive measurement program without being intrusive. One company attempted to measure media consumption and consumer purchases in the same household. Generally, participants had to subject themselves to extensive monitoring and extraordinary procedures to collect the data. In some ways, that made them unlike the general marketplace to which the results would be generalized. So the measuring had a tendency to get in the way of the natural consumer behavior. With the Internet, data collection is more transparent—web users are not really aware that their movements through the Web are being recorded and analyzed.

Cookies

The Internet has made it easy to track what consumers do, where they go, and what interests they have. One way of keeping track of that data is through the use of *cookies.* Webopedia defines cookies as a "message given to a Web browser by a Web server. The browser stores the message in a text file. The message is then sent back to the server each time the browser requests a page from the server." David Whalen made this analogy on www.cookiecentral.com:

> You drop something off [at the dry cleaners], and get a ticket. When you return with the ticket, you get your clothes back. If you don't have the ticket, then the [dry cleaner] man doesn't know which clothes are yours. In fact, he won't be able to tell whether you are there to pick up clothes, or a brand new customer. As such, the ticket is critical to maintaining state between you and the laundry man.

Webopedia goes on to explain: "The main purpose of cookies is to identify users and possibly prepare customized Web pages for them. When you enter a Web site using cookies, you may be asked to fill out a form providing such information as your name and interests. This information is packaged into a cookie and sent to your web browser which stores it for later use." So when you return to that same web site, your browser will send the cookie to the web server letting it know who you are—it's your ID card or your frequent shopper card. Then, the server can use this information to load up personalized web pages that may include content that interests you, based on information the site collected the last time you visited. So, for example, instead of seeing just a generic welcome page you might see a welcome page with your name and features on it. The use of cookies is frowned upon by privacy advocates but hailed by marketers and webmasters alike in its ability to offer customized information to visitors.

FIGURE 10.1

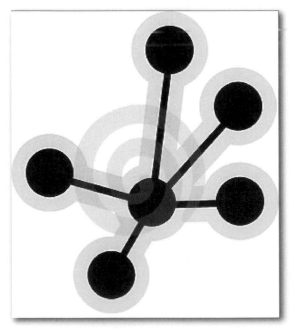

Collaborative Filters

Collaborative filtering software examines a user's past preferences and compares them with other users who have similar interests. When that user's interests are found to match another group of users, the system starts making suggestions of other things that members of this group like. In the article "Collaborative Filtering," author Francis Heylighen (2001) stated, "The main idea is to automate the process of 'word-of-mouth' by which people recommend products or services to one another. If you need to choose between a variety of options with which you do not have any experience, you will often rely on the opinions of others who do have such experience."

Suppose you normally never listened to jazz music, but you liked artists A, B, and C a lot. If numerous other people who don't normally listen to jazz also like A, B, and C, but also like band D, the system might suggest D to you and be relatively confident that you'll like it. Amazon. com uses the technology to recommend other products with its "people who bought this product also purchased these other products" feature. Unlike previous music sorting procedures that required judgments based on personal tastes and opinions, web-based collaborative filtering is usually a process developed through the input of consumers. So we now have consumers dictating which songs belong in which category and should sit beside each other on playlists. Web 2.0 has increased the user-generated process by allowing users to create and share their playlists with other members of the social networking world. Through services such as imeem and iLike, data is gathered when members group songs together into playlists. That cumulative information, combined with an individual user's profile and activities, helps the system make intelligent recommendations of additional music each user is likely to enjoy.

Collaborative filtering can help artists and their marketing analysts determine where an artist fits in relative to other artists, but only if the artist has a large enough following to be included on one or more collaborative filtering web sites. (See the section on determining the target market for your artist.)

GENERAL INFORMATION ON FINDING YOUR MARKET ONLINE

Author Frances Vincent, in her book *MySpace for Musicians*, brought up some good points for identifying your target market. Among those, Vincent suggested asking your fans directly; researching contemporaries and competitors; becoming a student of pop culture by listening, watching, reading, and going places; and networking.

One marketing article by Donna Gunter (2006) put it in these terms: "Are you fishing where the fish are?" Gunter went on to describe several research tools available online to help research the target market. Here is an expanded list of those sources:

1. *Professional associations.* They generally have information related to the market based on commissioned research studies. For the music business, that would include, but is not limited to, the National Academy of Recording Arts and Sciences (NARAS), the National Association of Recording Merchandisers (NARM), the Association of Independent Music (AIM), various genre-specific trade organizations, and the International Federation of Phonographic Industries (IFPI).
2. *Professional conferences.* Many of the trade associations hold annual conferences with panels and presentation on the latest research in consumer trends. Several notable conferences worth attending are South by Southwest, NARM, and MIDEM.
3. *Trade and consumer publications.* Read up on the market and the industry by subscribing to the top publications. For the music business that would include, but is not limited to, *Billboard, CMJ Network, Radio and Records, Pollstar, Hollywood Reporter, Rolling Stone* and *Variety*. Check out Billboard. biz. There are also genre-specific magazines such as *Source, Vibe, Country Weekly, Downbeat, Remix, Christian Musician,* and *Alternative Press.*

Table 10.2   Industry Organizations

Music Industry Trade Associations

National Association for Recording Merchandisers (NARM)	www.narm.com
National Academy of Recording Arts and Sciences (NARAS)	www.grammy.com
American Association of Independent Music	www.a2im.org
International Federation of Phonographic Industries (IFPI)	www.ifpi.org
Songwriters Guild of America (SGA)	www.songwriters.org
American Federation of Musicians	www.afm.org
Audio Engineering Society	www.aes.org
National Association of Recording Industry Professionals (NARIP)	www.narip.com
National Association of Broadcasters	www.nab.org
International Alliance for Women in Music	www.iawm.org

Source:http://musicnewsdaily.com/org.html.

Table 10.3   Industry Conferences

Examples of Industry Conferences

Billboard Magazine sponsored events	South by Southwest SXSW Music & Media Conference: Austin, Texas
The music industry's most powerful business-to-business events, including Digital Music Live, Mobile Entertainment Live, Billboard Music and Money Symposium, Billboard Touring Conference and Awards, R&B Hip Hop Conference and Awards, Latin Music Conference and Awards, and more. www.billboardevents.com/billboardevents/index.jsp	Showcases hundreds of musical acts from around the globe on more than 50 stages in downtown Austin. By day, conference registrants do business in the SXSW Trade Show in the Austin Convention Center and partake of a full agenda of informative, provocative panel discussions featuring hundreds of speakers of international stature. www.sxsw.com

Table 10.3   Industry Conferences—cont'd

Examples of Industry Conferences

Midem: Palais de Festivals, Cannes, France	Radio and Records Magazine R&R Convention
Nearly 10,000 music and technology professionals from more than 90 different countries, including delegates from the recording, publishing, live, digital, mobile, and branding sectors gather to do deals, network, learn, and check out new talent. www.midem.com	Attracting the top broadcast and recording-industry executives from around the country, the R&R Convention is recognized as the premier annual conference for the radio and record industries. www.radioandrecords.com/Conventions/RRconvention.asp
Millennium Music Conference	**Audio Engineering Society**
For over a decade, the Millennium Music Conference has educated emerging talent, showcased new music and entertained the community. Music industry professionals attend as panelists, speakers, mentors, exhibitors, and talent scouts network, do business, and share their experience with musicians, registrants, and attendees. The community joins in the celebration of emerging talent, independent artists, and new music. http://musicconference.net/mmc12	Conventions are held annually in the United States and Europe. Each convention has valuable educational opportunities, including a full program of technical papers, seminars, and workshops covering current research and new concepts and applications. An integral part of each convention is a comprehensive exhibit of professional equipment. www.aes.org/events

National Association of Recording Merchandisers

The NARM Annual Convention & Marketplace is considered the industry's premier venue to learn, discuss, and meet. Industry players come together to make business deals, hear live music, get the latest research, see the most up-to-date technology, showcase new product lines, hammer out solutions to industry issues, and network to find new business partners. www.narm.com/esam/AM/Template.cfm?Section=Convention

4. *Online discussion forums/lists.* Go for the industry-oriented forums rather than fan-oriented sites. Entering terms such as "music business forums" into a search engine should return a plethora of sites. For example, www. starpolish.com and www.getsigned.com are good industry reference sites. Also try AllMusic.com and music.AOL.com.

5. *Online networking.* Find out where other industry folks hang out online and get to know them. Or visit their web site and look for a "contact us" link. Drop them an e-mail to encourage online discussion.

6. *Blogs.* Leaving comments for other bloggers in the industry can help raise your profile in addition to fostering intellectual conversation and learning from the bloggers. Don't hesitate to leave questions in your blog comments—they may be answered by other readers as well as the blogger. Check out AOLMusicNewsBlog.com and Billboard.blogs.com.

7. *E-zines.* The use of e-zines to promote an artist is covered in the chapter on Internet promotion, but these e-zines can also be a good place to gain an understanding of the target market. If the publication is professional and successful, the publishers probably already know a great deal about their readership. Although they may not share that information willingly or without cost, the results of their knowledge and understanding should be evident in the subject matter of the e-zine. (For example, if the e-zine has an article on the quality of MP3 files compared to other formats, then perhaps the readership is technology oriented.) Gunter noted that reading these e-zines will "give you a great overview of typical issues and problems faced by your target market." Check out www.ezine-dir.com/Music.

8. *The radio industry.* Radio has a long history of studying music listening trends. Radio stations routinely conduct marketing research. RadioandRecords.com has a section on ratings that lists the top radio markets and some format and demographic details on each.

Make It Easier for Your Market to Find You

In the early 1990s, Internet researchers alluded to the coming age of *broadcatch,* moving away from the days of broad*cast,* where multitudes of people consumed a common message or programming simultaneously. Broad*catch* on the other hand, refers to the concept of media consumers seeking out programming, entertainment, or information. Under this scenario, the role of marketing and advertising shifts from that of "carnival barker" to that of information provider—being there with the product information when the consumer is in the market and looking. With that comes the shifting role of marketing to making it easier for the market to find you rather than you finding your market. Part of that effort is outlined in the chapter on optimizing your web site—the idea of being at the top of search results for the product category. For this effort to be successful, it is necessary to think like the consumers and gain an understanding of the process *they* go through to find *you.* What search terms are they likely to use? Do they always correctly spell the words? What if your band is an alternate spelling of a commonly known word, such as Boyz instead of Boys. Will your potential fans know to spell Boyz with a "z?" If not, all keyword-based advertising and search engine optimization should

FIGURE 10.2

include spellings and keywords most likely to be used *by members of the target market.* That is when marketers regret it if they have to use a domain name with a hyphen, or an extension other than dot-com, because most potential customers may not remember the subtle differences.

Think Virtually, Act Locally

Be sure to find and post notices at every possible local site where your target market is known to hang out—just like sniping (putting up posters) in the non-virtual world. If there are local web sites that post upcoming events and concerts, make sure to find these sites for each geographic market on the tour schedule and post a notice of the event, complete with viral components directing the visitor to the artist's web site. Potential fans may be perusing these sites looking for things to do locally and may have no prior awareness of your artist. Again, this requires getting into the mind of the potential consumer and imagining the process they may go through in discovering your artist.

DETERMINING THE TARGET MARKET FOR YOUR ARTIST

Established artists with a consumer base have the advantage when it comes to identifying the artist's target market. Generally these fans have been through some type of marketing research process as they were buying tickets or music, attending concerts, or just using the artist's web site. All of those fan interactions provide the opportunity for marketers to gather information about the fans to improve the marketing efforts. Established artists are more likely to appear on collaborative filtering web sites such as imeem, where a bit of research can determine the extent of overlap in fan base with other artists. Go through all the music-oriented sites that use collaborative filtering to recommend music and search for mentions of your artist. When they appear, go to the profile pages for these fans and get to know them. See what other music they are listening to, and make some inferences from that feedback.

It's generally more difficult to identify the target market for a new artist. Without a sales track record, the marketing department of a record label must make some assumptions about a new artist. Marketers would be wise to examine market characteristics that relate to other artists who are perceived as similar or who would appeal to a similar market. They may look at consumer information for more established acts and decide that this is the market they should go for. And at times, the product itself, in this case the artist, may undergo some modifications to successfully appeal to the "target market." It is also possible to survey visitors online or at live shows regardless of size of an artist's fan base.

This is also an area where online marketing and traditional marketing need to overlap. Data gathered in the real world can be applied to the virtual world and vice versa. At clubs and venues where the artist appears regularly, analyze the other musical acts that perform there to determine if there is any audience overlap. (It is the same principle as collaborative filtering, only performed in a more casual way.)

SPEAK TO YOUR MARKET

One of the most important aspects of successful marketing is in knowing *how* to communicate with your market. It is not enough to know where they are and how to find them, but you must also understand how to *reach* them with your marketing message. Certain market segments, particularly teenagers and young adults, will reject marketing messages from companies they perceive as not understanding them. So the message itself becomes important, and not just the words, but the look and feel of the web site, the logo, the images, and so on. These design elements and messages must be fine-tuned for the market, in the vernacular that the target market is accustomed to hearing.

In his book *How to Promote Music Successfully on the Internet,* author and musician David Nevue introduced a concept he calls "targeting by site." The idea is that a certain segment of *your market* may be looking for or respond to something in particular or may respond to something of a more general nature. For example, if your artist is a blues slide guitar player from the Panhandle of Florida, perhaps you should design a web site around North Florida Blues in an attempt to attract anyone who is interested in that genre in that area of the country.

Landing Pages

Nevue even suggested creating particular topic-oriented web pages and submitting them separately to search engines. For record labels that represent more than one genre of music, the label would do well to feature more than one web site. The label would also want to create specific pages that feature each of the artists who are currently the top priority in marketing. These pages are referred to as *landing pages.* Landing pages are specifically designed pages that are intended to be the page that the web visitor "lands on" when looking for something in particular and are directed to your site. This page would be more finely tuned to appeal to the segment of your market that is looking for something specifically and has found you through the use of a particular set of keywords that indicate their interest. In an article titled "Creating Effective Landing Pages" on TamingtheBeast.net, author Michael Bloch described landing pages this way:

> In marketing terms, it's a specialized page that visitors are directed to once they've clicked on a link, usually from an outside source such as a Pay Per Click ad. The page is usually tightly focused on a particular product or service with the aim of getting the visitor to buy or take some form of action rapidly that will ultimately lead to a sale.
>
> **(www.tamingthebeast.net/articles5/landing-pages.htm)**

Bloch explained that too many web site marketers labor under the assumption that the product purchase scenario goes like this: The customer arrives at the company's home page, the customer then selects an option from a menu or from an offer on the page, the customer clicks to the page with the product he or she is interested in, and then the customer buys. But it often does not work that way.

Landing pages are particularly useful to multiartist sites like nashvilleim.com. John Haring of Nashville Independent Music.com stated, "On nashvilleim. com, every artist has their own unique landing page address and is given custom MySpace banners and web site links to make the landing page almost like the home page of their own web site. In fact, some artists use it, along with MySpace, as their only web site."

The landing page should be the first page that customers are directed to when they have found your product and your web site through search engine keywords, through online advertising, or in response to your e-mail touting the particular product. At that point, the customers know what they want, the seller knows what the customers want, and the seller should deliver the customers to the product as quickly as possible. One analogy is that if a customer wanted to buy a pair of tennis shoes and the retailer had the ability to magically transport the customer to the correct shoe rack in the correct store, why would the retailer drop the customer off at the front entrance of the mall?

Use different landing pages to test different offers and creative treatments. In his article "How to Write an Effective Landing Page," Ivan Levison suggested, "You can test variables by sending prospects to unique landing pages. Just measure the click through rate and you'll find out fast what works best." The web offers companies a unique opportunity to target customers by creating alter ego web pages and web sites that appeal to each segment of the market. By using landing pages, a particular customer may never be aware of the variety of other markets this company serves.

CONCLUSION

The first step in developing a marketing plan is identifying and learning about your target market. The more you learn about that market, the more effective the marketing plan can be. It is important to reach out to customers with a communication strategy that speaks to them and makes them feel like the company or brand understands them. On the Internet, much of that involves being easy to find. To do that, you must get inside the consumer's head and go through the scenario the consumer will enact when shopping for and deciding on your product. How do customers go about finding you in the vastness of the Internet? What techniques do they use? How does your fan base find out about new music and new artists? How loyal is your fan base? Is it necessary to constantly find new customers, or do previous customers keep coming back for more?

Then, once the customers have found you, it is important to give them the right information to help them make that purchase decision. What motivates them to buy? Are they driven by costs, quality, uniqueness, a bond with the product? Major corporations spend millions of dollars on research to better understand their market and can modify not only the marketing messages but sometimes the product itself, and certainly the web site.

GLOSSARY

Behavioral targeting – Creating market segments based on what Internet users purchase online and what types of sites they visit.

Broadcatch – The concept of media consumers seeking out and controlling their consumption of programming, entertainment, or information rather than being passive consumers.

Collaborative filtering – Examines a user's past preferences and compares them with other users who have similar interests. When that user's interests are found to match another group of users, the system starts making suggestions of other things this person may like.

Cookies – Parcels of text sent by a server to a web browser and then sent back unchanged by the browser each time it accesses that server. The main purpose of cookies is to identify users and possibly prepare customized web pages for them.

Landing pages – Also known as jump pages. It is the particular page a customer is directed to by a link based on what that customer is looking for. The page is usually tightly focused on a particular product or service with the aim of getting the visitor to buy.

Market segmentation – Breaking a total market down into groups of customers or potential customers who have something significant in common in terms of their needs and wants or characteristics.

Target market – A segment of a specific market that your company has identified as your customers or clients.

REFERENCES AND FURTHER READING

Bloch, Michael. Creating effective landing pages, www.tamingthebeast.net/articles5/landing-pages.htm.

Gunter, Donna. (2006). How to find your target market online, www.amazines.com/article_detail.cfm/124420?articleid=124420.

Heylighen, F. (2001, January 31). Collaborative filtering. *Principia Cybernetica*, http://pespmc1.vub.ac.be/COLLFILT.html.

Jesdanun, Anick. (2007, December 1). Ad targeting grows as sites amass data on web surfing habits. The *Tennessean* daily newspaper.

Levison, Ivan. How to write an effective landing page, www.businessknowhow.com/internet/landing-page.htm.

Nevue, David. (2005). How to promote your music successfully on the Internet, www.promoteyourmusic.com.

Vincent, Frances. (2007). *MySpace for Musicians*. Thomson Course Technology, Boston, MA.

Whalen, David. (2002, June 08). FAQs about cookies, www.cookiecentral.com.

www.webopedia.com.

CHAPTER 11

Successful Promotion on the Web

INTRODUCTION

From a business standpoint, it wouldn't make any sense to open a restaurant—to put all that effort into designing and building a fine establishment—without a strategy to promote the restaurant to potential customers. But that is what people do when they spend their entire Internet budget and effort on building a perfect web site with no plans to promote it.

In the restaurant analogy, management must make certain obvious marketing efforts to become successful, including advertising, getting reviews, creating word of mouth, and offering coupons, to name a few. There is a common expression: "No one wants to eat at a restaurant with no cars parked out front." The presence or absence of a full parking lot indicates a restaurant's popularity and level of success. Online, search engine ranking achieves the same purpose: a high search engine ranking denotes that your web site is one of the more popular sites in its particular keyword-driven category, and as a result, more Internet users are likely to check it out. But how does one achieve this high ranking? Restaurants have been known to have employees park their cars out front on slow nights to give the appearance of a popular spot. Web sites use search engine optimization (see the section on search engines) to achieve this same result.

The restaurant would also need to advertise to create awareness, establish the brand, and bring customers in the door. Likewise, a web site developer will need to buy banner ads and use Google's or Yahoo!'s advertising programs. The restaurant would actively seek out restaurant reviewers, travel guides, and other opinion leaders in hopes of spreading the word through media outlets. Ditto for the artist's web site, where the Internet marketer will attempt to get bloggers to write about the artist and the site and will try to get online magazines to publish features about the artist.

The successful restaurant will employ techniques to reward customer loyalty and repeat visits, including frequent customer promotions (buy 10, get the next one free), drawings for free meals, and other promotional incentives for repeat customers. The successful web site should offer similar incentives to encourage visitors to return to the site periodically. And certainly, word of mouth is one of the best forms of promotion and one that successful marketers cultivate by using subtle techniques to encourage satisfied customers to spread the word.

This chapter and the next will examine the online promotional tools necessary to turn a good web site into a popular, successful, and profitable web site.

BASICS FOR INTERNET PROMOTION

The Internet has opened new opportunities for musicians and small businesses to easily and economically promote their products on a global scale. Web marketing should be a part of every recording artist's marketing plan, but it should not be the only aspect of the plan. Even though the Internet has become a great tool for selling music, the traditional methods of live performance, radio airplay, advertising, and publicity are still significant aspects of marketing and should not be neglected.

Rule 1: Don't make the Internet your entire marketing strategy. Internet marketing should not be a substitute for traditional promotion. The two strategies should work together, creating synergy.

In the music business it is necessary to build brand awareness. Whether it's with the artist or the record label, you need to create a sense of familiarity in the consumer's mind. This can be done through many tactics available both on and off the Internet, which should be designed to lure customers to the artist's web site or drive them into the stores to buy the artist's records.

Although the web is a great tool to reach a large number of people with minimum expense, it is so vast that unless potential customers are looking for a particular artist, they are unlikely to stumble across an artist's site by chance. Also, many customers are more responsive to traditional marketing methods such as radio airplay and retail store displays. An artist's web site is a great place for customers to learn more about the artist's products—live shows and recordings—but it is not necessarily the best way to *introduce* new customers to the artist's products.

Rule 2: Build a good web site, but don't expect customers to automatically find it on their own.

A solid marketing plan incorporates the company web site into every aspect of marketing and promotion. And while the web site will be the cornerstone of Internet marketing efforts, it is just the start. Building a good web site is crucial to marketing success, but the old adage "If you build it, they will come" does not apply to the Internet.

Every marketing angle should tie in to the web presence. Internet marketing is more effective when it is conducted in conjunction with other aspects of the plan. Every piece of promotional material should contain the web site address to direct the potential customer to the web site. Any posters, flyers, postcards, or press releases should have the artist's web site address prominently displayed. And it goes without saying that CD tray cards should also contain the address. One band had hand stamps printed up at the local office supply store. The stamps had the band's web address on them, and the band requested that the bouncers use them to stamp customer's hands as they entered the venue where the act was performing.

Rule 3: Incorporate the artist's web site address into everything you do online and offline.

Now that we have established the relative function of Internet promotion, we will next examine the various structures for promoting one's brand and products on the Web.

SEARCH ENGINES

Search engines are online directories of web sites and web pages that web visitors use to find a topic of interest or a specific site. They help organize the Web so that visitors can make some sense out of the vastness of the Internet (Dawes and Sweeney, 2000). One of the goals when setting up your web site is to ensure its prominence in search engine results so that potential customers can find it.

It is gratifying to type your artist's name into a search engine and have their official web site listed as the first result. That is likely to happen for well-established artists, but it is not as common with lesser-known acts, especially if the band or artist's name is a common one or is associated with another product. Even if the site does come up as one of the search results, the web visitor will need to have knowledge of the artist and the intent to seek out that particular artist online. What about potential fans who are not yet familiar with the artist? How do you reach them? And how do you improve your standing in search engine results when the person doing the search is using more vague terms, such as "blues music" or "female blues vocalists in Atlanta."

As of press time for this book, the most popular search engines are those shown in Figure 11.1.

Nielsen found that Google had 55.2% of the search engine market in May 2007. In another study, the online analyst service Hitwise found that Google accounted for 64% of all U.S. searches in August of 2007, with Yahoo! capturing 23%. By February 2008, Google was up to 66% and Yahoo! had 20%.

The various popular search engines each have their own way of categorizing and prioritizing their listings and search results. It is wise to understand how each finds and categorizes new web pages and then proceed with establishing your presence on each. Even though search engines are likely to find your site and list it, you should still submit a request to be listed and be prepared to list the relevant keywords and subject categories that apply to your site. Some of these

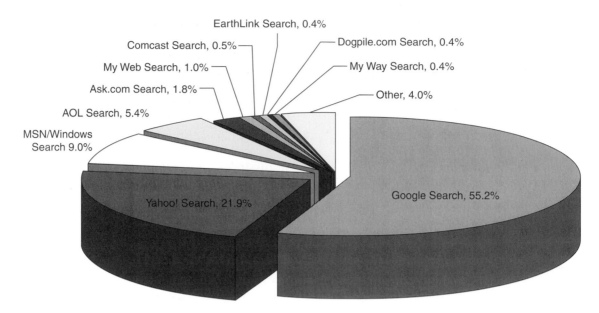

Search Engine Popularity - Share of Searches

EarthLink Search, 0.4%

Comcast Search, 0.5%

Dogpile.com Search, 0.4%

My Web Search, 1.0%

My Way Search, 0.4%

Ask.com Search, 1.8%

Other, 4.0%

AOL Search, 5.4%

MSN/Windows Search 9.0%

Yahoo! Search, 21.9%

Google Search, 55.2%

Source: Nielsen/NetRatings Megaview Search, May 2007

FIGURE 11.1
Search engine popularity.

search engines use *spiders*, software programs that scour the Internet and analyze web pages. Others use humans, or a combination of both.

Numerous commercial services offer *search engine optimization*. Webopedia defines search engine optimization as "the process of increasing the amount of visitors to a web site by ranking high in the search results of a search engine. The higher a web site ranks in the results of a search, the greater the chance that that site will be visited by a user" (www.Webopedia.com). Most search engines rank their results based on several factors. By paying attention to these factors, you can improve the chances of getting a top listing:

1. *Keywords.* Choose *keywords* that are targeted and represent the web site's topic accurately. These keywords should appear in the page text in addition to being listed in the meta tag keyword section. Many search engines rely on page text to confirm that the site's topic is accurate and corresponds to the keywords. If certain words have alternative spellings or are commonly misspelled, the variations should be listed in the meta tag section because these misspellings will not be visible to the web visitor. Keywords should also appear high on the page, within the first 100 characters. When ranking results, search engines give higher priority to keywords that appear high on the page, in the title, in the description,

in the URL, in headings, in the ALT tag for graphics,[1] and in the link text for inbound links (Monash, 2004).

2. *Content.* Make sure the content of the site is represented accurately. Search engines monitor traffic to your site. If web visitors quickly leave your site because it is not what they expected, search engines will know that the subject matter is not what the visitor is looking for based on the search terms they entered (Walker, 2006). The search engine may then lower the ranking if many search engine users fail to click through and explore your site.

3. *Links.* The number and quality of inbound links helps the search engine spiders determine the validity and popularity of your site. Link popularity is one factor the search engines gauge when determining ranking. If many people link to your site, then it must be one of the more popular sites. Danny Sullivan on SearchEngineWatch.com has suggested the following:

> Go to the major search engines. Search for your target keywords. Look at the pages that appear in the top results. Now visit those pages and ask the site owners if they will link to you. Not everyone will, especially sites that are extremely competitive with yours. However, there will be noncompetitive sites that will link to you—especially if you offer to link back.
>
> **(Sullivan, 2007)**

There are three ways to get a web site listed in a search engine: (1) submit the site directly to the search engine via an URL submission form, (2) wait for the search engine to find the site, (3) pay the search engine to index the site. Sometimes the submission form is buried deep in a search engine's site; for Google, you must click the link "about Google" and then select "submit your content to Google" from a menu. Yahoo! has a link at the bottom of the page that says, "Suggest a site." More information is available on search engine optimization in Chapter 7.

E-ZINES

Electronic or online magazines offer a good way to introduce your target market to your act. MarketingTerms.com defines *e-zines* as an electronic magazine, whether posted via a web site or sent as an e-mail newsletter. (The term is loosely used to describe e-mail newsletters, albeit without the magazine formatting; see section on using e-mail for promotions.) Short for electronic magazine or fanzine, some are electronic versions of existing print magazines, whereas others exist only in the digital format. The web-posted versions usually contain a stylized mixture of content including photos, articles, ads, links, and headlines, formatted much like a print equivalent. Most e-zines are advertiser-supported, but a few charge a subscription.

Many established music e-zines are genre specific or have particular subject areas dedicated to genres. They may feature music news, concert and album reviews,

---

[1] The "alt" tag is an HTML tag that tells browsers to insert a description in the place of a missing image or when the cursor is scrolled over the image.

interviews, blogs, photos, tour information, and release dates. As a result, their readers are predisposed to be receptive to new and unfamiliar artists and their music, provided that the artist is within the genre that the e-zine represents. A study on readers of the Americana music magazine *No Depression* found that 90% of their readers found out about new music from an article, one published either in a print magazine or in an online version.

What to Send

E-zines are mostly interested in feature articles and press releases pertaining to some newsworthy item (such as an album release or a tour schedule announcement). A feature story should include biographical information as well as the newsworthy information—in other words, it is part bio and part press release. It helps if you write the story as if it were going to appear unedited in the online publication, in the inverted pyramid style. Always include a publicity photo or two along with the article for submission. And *always* include plenty of links to specific pages on your web site that pertain to the featured topics (tour page for touring news, product page for record release news, etc.).

Do not send out a press release if you have nothing that is considered newsworthy. A redesign of your web site might be interesting to fans who have signed up for e-mails, but to the casual reader who was probably unfamiliar with your old web site, it's not news and doesn't warrant a press release. In his book *How to Promote Your Music Successfully on the Internet*, David Nevue emphasized that before you send out a press release announcing your new CD, ask yourself "who cares?" and let the answer to that decision determine if, and to whom, you should distribute that information.

Where to Send It

The Ezine Directory has a listing of many of the better-known music e-zines, along with descriptions and ratings of each (www.ezine-dir.com/Music). The goal is to find those with the correct target market and submit articles, music, and photographs to the editor, encouraging him or her to include a link to the artist's web site. Some e-zines have submission forms available on their web site, whereas others are not as specific about their submission policy. But don't give up; Dawes and Sweeney (2000) suggest sending an e-mail to the editor to inquire about the potential to have your artist featured in the publication. Provide editors with writeups to make their job easier. When these reviews and articles appear, place a link on the artist's site directed to the specific page on the e-zine site that includes the article. To find a list of e-zines, try one of the directories, such as the ultimate band list or www. ezine-dir.com.

The Music Industry News Network (www.mi2n.com) features articles and news of various independent artists and will accept submissions for news items. On the home page, click "submit your news" from the menu bar at the top and then fill out the form. The submission form allows you to include your web site address, a link to sound files, and a link to a graphic file. For a fee (currently $69), their PR Syndicate service will send out your press release to major music news sites, including the appropriate "groups" at Google and Yahoo! and their related *Music Dish Network* sections of major social networking sites.

CDBaby.org has a feature called "ADD news!" that gives musicians (CD Baby members) a chance to post news items. Submissions appear on a bulletin-board style page with the most recent stories at the top. For submissions to this outlet, it is strongly advised that you use carefully selected headlines and a brief overview of the story, because only the first few sentences appear in the posting. Readers must click to read the full story. So make the headline and story lead compelling.

BeatWire.com is a web site dedicated to press release distribution for independent musicians and record labels. The cost is $149 and includes distribution to all the major music publications and media outlets, including monthly music magazines, college radio stations, and weekly and daily newspapers in hundreds of markets. This would be a good idea for an artist who is ready to move to the next level and has a compelling news release that is likely to be picked up by these national and regional publications.

It might be best to first make an electronic press kit (EPK) consisting of music, photos, videos, bio, discography, tour dates, contact information, and other elements found in a standard press kit. Sonicbids is a web site that creates and stores EPKs for musicians. The site states, "Your EPK does not replace your web site. It replaces the costly physical press kits you put together to get booked at festivals and venues or to secure other performance and licensing gigs" (www.sonicbids.com).

Resources for E-zines and Distribution of Press Releases

The Ultimate Band List: www.ubl.com

CD Baby: www.cdbaby.org

BeatWire: www.beatwire.com

The Ezine Directory: www.ezine-dir.com

Music Industry News Network: www.mi2n.com

MusicDish: www.musicdish.com (try the open review and "submit your article")

John Labovitz's E-ZINE LIST: www.e-zine-list.com/titles_by_keyword/music/
    page1.shtml

Zinester Ezine Directory: www.zinester.com/Music-

Dmusic: http://news.dmusic.com/submit

PRWeb: www.prweb.com

Updated information available at www.WM4MB.com.

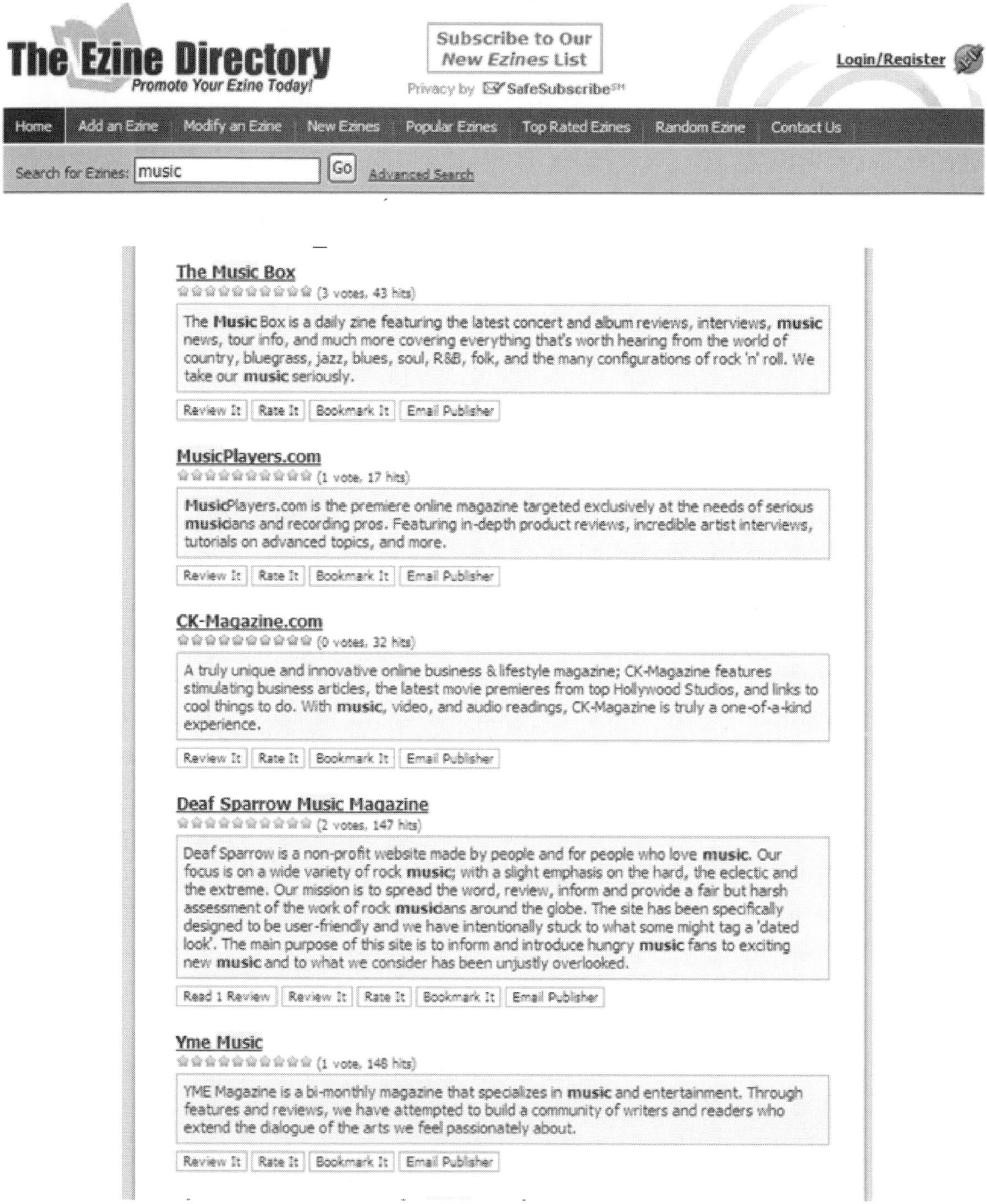

The Ezine Directory
Promote Your Ezine Today!

Subscribe to Our
New Ezines List

Privacy by ☑️ SafeSubscribe℠

Login/Register

| Home | Add an Ezine | Modify an Ezine | New Ezines | Popular Ezines | Top Rated Ezines | Random Ezine | Contact Us |

Search for Ezines: `music` [Go] Advanced Search

The Music Box
☆☆☆☆☆☆☆☆☆☆ (3 votes, 43 hits)

The **Music** Box is a daily zine featuring the latest concert and album reviews, interviews, **music** news, tour info, and much more covering everything that's worth hearing from the world of country, bluegrass, jazz, blues, soul, R&B, folk, and the many configurations of rock 'n' roll. We take our **music** seriously.

[Review It] [Rate It] [Bookmark It] [Email Publisher]

MusicPlayers.com
☆☆☆☆☆☆☆☆☆☆ (1 vote, 17 hits)

MusicPlayers.com is the premiere online magazine targeted exclusively at the needs of serious **music**ians and recording pros. Featuring in-depth product reviews, incredible artist interviews, tutorials on advanced topics, and more.

[Review It] [Rate It] [Bookmark It] [Email Publisher]

CK-Magazine.com
☆☆☆☆☆☆☆☆☆☆ (0 votes, 32 hits)

A truly unique and innovative online business & lifestyle magazine; CK-Magazine features stimulating business articles, the latest movie premieres from top Hollywood Studios, and links to cool things to do. With **music**, video, and audio readings, CK-Magazine is truly a one-of-a-kind experience.

[Review It] [Rate It] [Bookmark It] [Email Publisher]

Deaf Sparrow Music Magazine
☆☆☆☆☆☆☆☆☆☆ (2 votes, 147 hits)

Deaf Sparrow is a non-profit website made by people and for people who love **music**. Our focus is on a wide variety of rock **music**; with a slight emphasis on the hard, the eclectic and the extreme. Our mission is to spread the word, review, inform and provide a fair but harsh assessment of the work of rock **music**ians around the globe. The site has been specifically designed to be user-friendly and we have intentionally stuck to what some might tag a 'dated look'. The main purpose of this site is to inform and introduce hungry **music** fans to exciting new **music** and to what we consider has been unjustly overlooked.

[Read 1 Review] [Review It] [Rate It] [Bookmark It] [Email Publisher]

Yme Music
☆☆☆☆☆☆☆☆☆☆ (1 vote, 148 hits)

YME Magazine is a bi-monthly magazine that specializes in **music** and entertainment. Through features and reviews, we have attempted to build a community of writers and readers who extend the dialogue of the arts we feel passionately about.

[Review It] [Rate It] [Bookmark It] [Email Publisher]

FIGURE 11.2
The Ezine Directory. (Courtesy of The Ezine Directory, www.ezine-dir.com.)

USING E-MAIL FOR PROMOTION

Perhaps one of the most successful marketing strategies brought about by the Internet is the use of e-mail to effectively target and promote commercial enterprises. The use of periodic mass e-mailing allows an artist to announce new recordings and live appearances, contact the press, disseminate news and accomplishments, and keep fans coming back to the web site. The effective use of e-mail requires that you develop a list of interested fans, keep the list updated, and correctly target portions of the list for appropriate e-mail announcements. Research has shown that e-mail is very effective in driving traffic to the web site, but only if the e-mail contains something of interest for the consumer.

Building and Managing Your List

The best possible way to develop a quality e-mail list is by requesting that fans and web visitors provide their e-mail address so that they can receive valuable updates, news, and perhaps savings opportunities. It is important to provide a sign up form in a prominent location on the web site. One way to motivate fans to provide their address is by offering something extra to those who become fan club "members" by providing their address. This can be anything from access to "restricted" areas of the web site, where they can receive free items such as audio tracks, to offering contest prizes to those who sign up.

While it is important to provide web site visitors with the opportunity to submit their e-mail addresses when they visit the web site, it is equally important to provide sign up opportunities at gigs and in retail settings. Always provide an e-mail sign up sheet at the bar or the front door, or have a worker pass around a sign-up sheet. It also helps to make an announcement from the stage asking fans to provide their e-mail addresses so they can receive valuable updates and information on future performances.

G-Lock software advises not to "generate e-mail addresses using special tools you may find on the Internet; don't harvest the e-mail addresses from web sites; and don't purchase a ready-made list." The best quality lists come from grassroots efforts to sign up those people who are interested in the artist and willing to accept e-mail newsletters.

It is important to inform people that by providing an e-mail address, they will be receiving periodic updates—in the form of an e-mail—about the artist. In this era of spamming, it is important to provide recipients of the e-mail newsletter with the opportunity to unsubscribe from the e-mail list. Spamming is the activity of sending out unsolicited commercial e-mails. It is the online equivalent of telemarketing. Fans who provide their e-mail addresses should be told that by providing this information, they will be receiving correspondence—the specifics of which should be spelled out. You should never sell or offer your e-mail lists to third parties. The addresses should be kept confidential and care should be taken when sending out bulk e-mails that the addresses of other recipients are not visible and available to any of the recipients. In the absence of specialized software for managing e-mail lists, you can always send the e-mail correspondence to yourself and *blind copy* (BCC) the message to those on the list. By using this commonly found e-mail function, the addresses of the other recipients are not visible in the message heading.

Some of the important factors in e-mail list management include the following:

1. Categorizing the list to send targeted e-mails only to appropriate groups. Touring information is only important to fans in the area of the performances. Leave the comprehensive touring schedule to the web site and use e-mail for only those fans within driving distance to the venue.
2. Keeping the list current by purging addresses that are no longer valid.
3. Promptly removing subscribers who asked to be removed.
4. Providing automated ways for new subscribers to be added to the list (there are many software programs that will provide this feature).

The available software programs, both web-based and computer-based, help manage e-mail lists and enable you to create subcategories. Some of the features to look for in e-mail management software include managing the list, mail merge, verifying e-mails, removing dead addresses, managing unsubscribe requests, and e-mail tracking. It is recommended that a web master invest in one of these programs if there is any volume of e-mail that cannot be handled promptly at all times. *Mail merge* is important for personalizing messages with the recipient's name in the salutation. An automatic responder is helpful for sending a confirmation to people who have responded to your e-mail or purchased your product.

THE E-MAIL NEWSLETTER

The e-mail newsletter, sent to willing recipients who have signed up to receive the newsletter, can be an extremely effective and inexpensive way to promote an artist. In his article "7 Reasons to Start an Email Newsletter Today," Rich Brooks extolled the virtues of utilizing this marketing technique, for the following reasons:

1. They complement your web site—a newsletter is the counterpart, the outreach portion of your online marketing plan, or, as Brooks stated, it is "like white wine to fish (not that your web site stinks like fish)." Most people spend a great deal of time reading and responding to e-mails, so this puts you right in front of their faces.
2. They are more cost effective than print newsletters. Forget about all those printing costs and postage!
3. They are interactive. You need to goad people into turning on their computers and typing in your web address (correctly, you hope). The fans read, they see a link of interest, they click on it.
4. You can track their effectiveness. "You can track which links in your newsletter are being clicked on and which are being ignored," stated Brooks.

5. They are viral. Word of mouth is easy because readers can forward the e-mail to friends they think will find it interesting.

6. You are preaching to the choir. Newsletters only go to fans who have signed up to receive them. These fans want to know where the next gig is or when the next recording is coming out.

7. You can start to build your fan base and subscriber base now to take advantage of the next generation of communication technologies.

Content and When to Send E-Mail Newsletters

Most news e-mails take a form adapted from print newsletters, with headlines and a brief overview of each topic, but with links to the full story on the web site for the added multimedia and graphic content. They should entice the reader to click on a link to get "the rest of the story" or view the item of interest. Be sure that each newsletter link directs the reader to the appropriate page of the main web site rather than the home page. Have a great opening, use short sentences, and focus on the recipient's self-interest. Newsletters can also contain fun items of interest, including stories, recipes, jokes, and links to sites of significance. The G-lock EasyMail web site advises users to not use all capital letters, don't write too long, correct misspellings and typos, and send a text version of the message along with the HTML part. The site also advises users to "excite curiosity in the subject line" to motivate the recipient to open and read the e-mail, to write as if you are conversing with another person, to keep it simple in format, and to place links at various intervals throughout the e-mail. The best e-mail newsletter is one that contains subject headlines with a teaser paragraph on each subject, providing enough information to get the point across but with links to more in-depth information on the artist's web site.

Much like a standard press release, e-mail newsletters should only be sent out when there is something new to report such as new music releases, announcements of upcoming touring, and award nominations. Newsletters that are sent out on a scheduled basis sometimes don't contain enough *news* to hold the reader's attention, and these senders often find that many of the recipients eventually request to be removed from the list.

Don't send out large files full of graphics and attachments. Instead, rely on the web site to provide the images. Although you may be tempted to send newsletters in the form of a PDF file, some recipients prefer HTML or text files and may find it difficult to download PDF files. The idea is to keep the e-mail correspondence as a small, easy-to-download file so that it is not filling up inboxes or being filtered by spam filters because of the active content or file attachments. Spyware, adware, and computer viruses (see Chapter 13) are often spread through the use of attachments, and many e-mail programs automatically filter them out.[2]

---

[2] Levels of filtering in one's e-mail program is usually consciously and proactively set by the user beginning with a default of no filtering on up to heavy filtering, then they work automatically.

SAUCE BOSS GAZETTE
Volume 16 No. 3 * April 1, 2008

RAW
IT'S HERE.
THE LATEST FROM THE SAUCE BOSS
It's called simply, "Raw". Unadorned, with little or no production, "Raw" is just three guys jamming in the studio. Majic John and Big Jim have never sounded better. The Sauce Boss scorching slide guitar slathered like habanero sauce over his very original blues. Sauce Boss uses no effects on his 53 Telecaster, his 59 Les Paul, or his "Bill Wharton Special". No need when you put them through a Fender Pro Amp made in 1948.
Listen to and ORDER RAW at CD Baby -
You can listen to samples of the entire CD at CD Baby.

COME HEAR THE NEW TUNES LIVE
April brings us to Miami, Ft. Lauderdale, Tampa, Lake Worth and Orlando. On Friday April 11, Sauce Boss will be on Ed Bell's live

If you are a radio DJ or a press writer and have not yet received your promo copy of "Raw", please send email to info@sauceboss.com. Blues Deluxe, a nationally syndicated radio show, with an audience of over 2 million, featured "Every Inch of Your Body" last week and Blues Matters, an international monthly magazine from the UK will feature a review in their April/May issue. Radio stations please send your playlists to info@sauceboss.com

OFFSHORE ALE COMPANY
Here's a little brewery on Martha's Vineyard that carries on the tradition of the old world pubs where food and brew is as comforting as coming home to Grandmas. (For more on our visit to Offshore Ale and Martha's Vineyard, visit sauceboss.com)

COCONUT COVERED ALMOND DATE CAROB WALNUT BALLS
Here's a dessert that's actually good for you. (And simple to make. For the recipe, visit sauceboss.com)

Quick Links...
Sauce Boss Website
Listen to Sauce Boss Podcasts
Join Our Friends on MySpace and Listen to some tunes
Sauce Boss Show Schedule
Watch Sauce Boss Videos on YouTube
Check out the new"RAW" CD

Email the Sauce Boss

Forward email

Email I

This email was sent to hutchtom@aol.com, by info@sauceboss.com
Update Profile/Email Address | Instant removal with SafeUnsubscribe™ | Privacy Policy.

Sauce Boss | 2873 Old Saint Augustine Road | Monticello | FL | 32344

FIGURE 11.3
Example of a Sauce Boss newsletter. (Reprinted by permission.)

You may find it necessary to ask your recipients to unblock your e-mail address so that your newsletters don't automatically end up caught in the spam filter. Ask your recipients to go to their spam folder, find your e-mail, and unblock the address—but, of course, this will be difficult to do if your only method of contact is the recipient's e-mail address. You can always use your personal e-mail address to contact members of your newsletter list who have inadvertently blocked your e-mails and notify them. It also might be a good idea to provide a notice when visitors sign up for your newsletter that will remind them to unblock your address.

Note that in the example of the Sauce Boss Gazette, the formatting is simplified for quick downloading of a small file—one that is less likely to fill up the recipient's mailbox. The use of bold caps for headlines and hot links is important for easy reading and to encourage readers to follow the links for more information on the official and other related web sites. Make sure that each link directs the reader to the specific page of the web site containing the material for that topic, rather than using a link to the home page (see landing pages in Chapter 10).

The e-mail newsletter, if used correctly and cautiously, can be an effective and inexpensive marketing tool that drives traffic to the web site, fans to the concerts, and increases recorded music sales. Rich Brooks, president of flyte new media stated, "an e-mail newsletter is the sharpest tool in your Web Marketing toolbox. If you're not sending one out you're missing a great opportunity to connect with your customers and prospects."

INTERNET RADIO

Internet radio—often referred to as *webcasting* and sometimes as streaming radio—is the concept of broadcasting much like a radio station, but via the Internet in lieu of a wireless signal. The obvious advantage over terrestrial radio is the absence of geographical limitations. Although the success of Internet radio has risen and fallen with the expenses associated with payments of performing rights royalties, it has consistently been a successful outlet for indie labels and emerging recording artists who do not always expect to be compensated for airplay of their music. Terrestrial radio airplay has been the key to selling recorded music since the days of early radio. It is still the most commonly used form of marketing new music to audiences and potential customers. But it is impossible for independent artists to gain much exposure through radio airplay on traditional terrestrial radio stations. Satellite radio, with its broader playlists and multitude of stations, has offered some opportunities for niche music, but Internet radio is the bastion of hope for frustrated emerging artists seeking exposure for their music. There are literally thousands of independent radio stations and Web 2.0 has created opportunities for anyone to create and manage their own online radio station with the playlist of their choice.

Music radio online generated 4.85 billion total listening hours in 2007, up 26.1% over 2006. Shoutcast (owned by AOL) remains the top platform/destination in the music radio segment, with 48.4% of total listening hours for the year, followed by Clear Channel Online, Yahoo! Music, AOL Radio Networks and Pandora.

AccuStream iMedia Research

Music podcasts are simply prerecorded shows, often in a radio-type format with narration and commentary intermingled with music. The major difference between podcasting and web radio is the time-shifting factor: podcasts can be downloaded, transferred to portable devices, and enjoyed at the convenience of the listener. Consideration must be given to the fact that portions of a podcast, including copyrighted music, may be copied and redistributed, whereas streaming radio is less likely to be captured and distributed illegally. This section concentrates on Internet streaming radio and getting airplay on web stations.

Popularity of Internet Radio

Arbitron reports that as of February 2007, an estimated 20% of the U.S. population 12 years of age and older listened to online radio in the past month, a total of 49 million people. And 29 million people 12 years of age and older, about

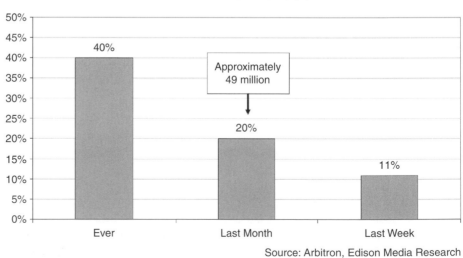

Source: Arbitron, Edison Media Research

FIGURE 11.4
Online radio listenership of U.S. population 12 years of age and older.

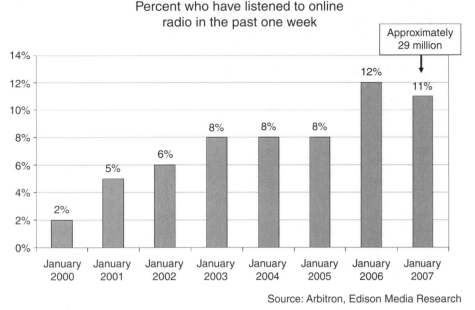

Percent who have listened to online
radio in the past one week

Approximately
29 million

Source: Arbitron, Edison Media Research

FIGURE 11.5
Weekly online radio audience over time, U.S. population 12 years of age and older.

11%, listened in the past week. In the article "The Day the (Web) Music Died," Robert Atkinson stated that as of July 2007, 60 million Americans were tuning in to web radio, surpassing many other Web 2.0 applications in popularity, including Facebook.

The popularity of online radio listening has been growing, but it seems to have leveled off in 2007. The 18- to 24-year-old group is most likely to listen to online radio, with half of all listeners under the age of 35.

Getting Airplay on Internet Radio Stations

There are three levels of webcasting stations on the Internet: (1) commercial stations that *simulcast* their programming, (2) professional entertainment entities that feature their own radio stations or podcasts, and (3) the plethora of small stations including those on services such as Live365 and SHOUTcast that allow music fans to operate their own stations.

SIMULCAST OF COMMERCIAL MEDIA STATIONS

Columnist John Taglieri commented that you can "literally travel around the globe… [and] still listen to your favorite radio station from home." Taglieri went on to state that while this is advantageous for listeners, it has not helped independent music because the emerging artist still cannot break into the playlist for the Internet version of these commercial stations because the programming is not unique to the Internet broadcast, but is the same program broadcast commercially in the station's geographic area. The advantage for major artists is that exposure is no longer limited to the geographic area of the station, and airplay on these stations could have an impact globally.

In his article "Radio Airplay 101: Traditional Radio vs. the Web," radio promoter Bryan Farrish stated that in the future, it will be just as difficult to get airplay on a big web radio station as it is to get airplay on a popular terrestrial station now. Farrish stated that the competition will heat up for slots on the most popular web stations, with artists and labels competing for the coveted airplay slots as they do now on FM radio. He added that as more sales come through downloads, the money saved on manufacturing will shift to promotion. Farrish concluded that "the amount of work it takes to get your songs heard will always be directly proportional to how many listeners you are trying to reach."

Clear Channel has created a web community called New! that allows unsigned bands the opportunity to upload their music to the web site. The songs are then offered to music fans on the web site for streaming and made available to Clear Channel's terrestrial program directors.

PROFESSIONAL ENTERTAINMENT ENTITIES THAT FEATURE WEBCASTING

Many media and music companies looking for additional ways to disseminate content and attract visitors to their web sites have turned to web radio as an economical way to achieve these goals. The most common way to provide this content is through podcasts that can be downloaded and consumed at the listener's convenience, but many who want to provide a real-time listening experience are turning to webcasting. For example, the American Society of Composers, Authors and Publishers (ASCAP) provides a network of streaming radio stations to increase exposure for ASCAP songwriters. The web site states:

> With eight channels of streaming audio, plus on-demand listening, video and podcasts, visitors to the ASCAP Network can discover and experience a diverse and dynamic community of songwriters, composers and publishers. In addition, ASCAP's highly-acclaimed Audio Portraits series gives listeners unique insight into the creative process, as told by the writers themselves.

(www.ascap.com/network/about.html)

Another example is pop icon Jimmy Buffet, who sponsors a streaming radio station at www.radiomargaritaville.com. The station started as webcast only in 1998 but was picked up by Sirius in 2005 and is now simulcast (Deitz, 2005).

For both media-sponsored webcasts and podcasts, getting on the playlist requires having some connection to the sponsoring organization, whether it's through membership or just targeting the right webcast sites and submitting appropriate material. David Nevue suggested identifying stations that take outside material

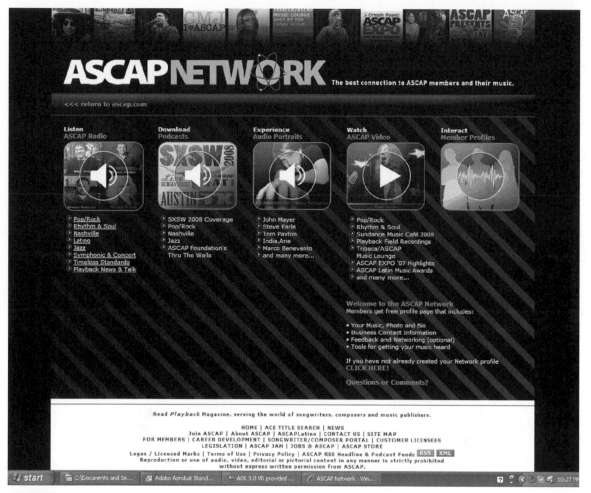

FIGURE 11.6
ASCAP Streaming Radio Network.

and match the genre, and then contacting them with a request to submit songs for consideration. Nevue stated, "some broadcasts receive a number of CDs to review, so it may take several weeks for them to get to yours" (p. 18).

PLETHORA OF SMALL STATIONS

The smaller, less popular web stations are easier to break into if they are targeted correctly. It is necessary to identify those stations for which your music is appropriate. In his article "Web Radio Stations and Getting Played," Bobby Borg (2005) wrote that the submission policy for these small stations is usually simple and involves e-mailing a music file to the webcaster. These small stations usually lack the exposure of larger, more popular stations, but getting airplay on these stations is inexpensive, and as more stations pick up the songs, the audience is cumulative. Borg stated, "SOME exposure is better than NO exposure—especially if it leads to a listener buying your record or coming out to one of your shows." Many of these amateur stations can be found on the services Live365.com and SHOUTcast. SHOUTcast provides listenership data for all stations for a 30-day period under the "stats" menu item.

To find appropriate stations on Live365, select the genre and from the list, select the appropriate stations (a broadcaster profile and broadcast schedule is provided), and contact the moderator of each. A free membership account is required to gain e-mail access. If your inquiry receives a favorable response, e-mail an MP3 file of the song you would like considered for broadcast. By all means, keep track of the stations on which you are receiving airplay. It helps to then promote the stations that support your music by including a link to them on your artist web site. To create your own station on Live365, you must pay a fee of $9.95 per month, which includes 150 MB of file storage and allows for up to 25 simultaneous listeners. SHOUTcast is free but requires that you use your own server and download the appropriate software. To request airplay, you must select the appropriate genre and scroll through the stations. A link to each station will send you to that station's site. Submission policies may vary.

Licensing and the Future of Webcasting

In the article "The Day the (Web) Music Died," Robert Atkinson stated that mandatory broadcast licensing fees threaten to end the enormous popularity of web radio. He stated that "Internet radio is a classic Web 2.0 application, offering diverse programming that caters to a range of specific groups." Whereas terrestrial broadcasters are required to pay royalties to performing rights organizations to cover licensing of the musical compositions for the songs they play, they are exempt from paying royalties for the actual sound recordings—money that would go to the label and artists. But for webcasting and satellite broadcasting, this additional royalty must be paid, usually to SoundExchange. SoundExchange's web site describes the organization as follows:

> SoundExchange is an independent, nonprofit performance rights organization that is designated by the U.S. Copyright Office to collect and distribute digital performance royalties for featured recording artists and sound recording copyright owners (usually a record label) when their sound recordings are performed on digital cable and satellite television music, Internet and satellite radio (such as XM and Sirius). SoundExchange currently represents over 3,000 record labels and over 20,000 artists and whose members include both signed and unsigned recording artists; small, medium and large independent record companies; and major label groups and artist-owned labels.
>
> **(www.soundexchange.com)**

Before 1995, there was no performance right for copyright holders of sound recordings in the United States. Other countries have been collecting royalties for sound recordings for years, providing revenue to recording artists and record labels for commercial airplay. SoundExchange goes on to state:

> The Digital Performance in Sound Recordings Act of 1995 and the Digital Millennium Copyright Act of 1998 changed that by granting a performance right in sound recordings. As a result, copyright law now requires that users of music pay the copyright owner of the sound recording for the public performance of that music via certain digital transmissions.
>
> **(www.soundexchange.com)**

Since then, there has been much debate over what the royalty rate should be based on and what is deemed fair to all parties involved. Rates were set to increase on June 15, 2007, to $0.0008 per performance: defined as streaming one song to one listener. Small broadcasters claimed that the new rate was too cost prohibitive; they banded together to protest the new rate, which is also set to increase 30% per year. In response, in August of 2007 SoundExchange offered a new rate to small broadcasters, defined as those earning $1.25 million or less in total revenues. Small broadcasters would pay a royalty rate of 10% or 12% of revenue, but only for SoundExchange member labels, and only if the small broadcasters signed on before September 14 of that year. This agreement would be binding through 2010, allowing these broadcasters to conduct business without concerns that rates would change before that date. The debate continues, and updates can be found at www.WebMarketingForTheMusicBusiness.com or www.WM4MB.com.

Resources for Internet Radio Promotion

Rhapsody: www.Rhapsody.com

SomaFM: http://somafm.com

Pandora: www.pandora.com

Live365: www.live365.com

SHOUTcast: www.shoutcast.com

PureVolume: www.purevolume.com

Digitally Imported: http://www.di.fm

International Webcasting Association: www.webcasters.org

Radio and Internet Newsletter: www.kurthanson.com

New!: http://www.clearchannelmusic.com/cc-common/artist_submission/index.htm?gen=2

GRASSROOTS

Grassroots marketing is defined as unconventional street-level marketing using word-of-mouth influence and that of opinion leaders to disseminate a marketing message among potential customers. Eschewing the top-down approach, grassroots marketing bypasses the traditional marketing methods that rely on expensive, slick advertising messages delivered through traditional media vehicles.

Guerilla marketing is another term closely associated with grassroots marketing. Whereas grassroots marketing refers to a street-level, door-to-door campaign of peer-to-peer (P2P) selling, guerilla marketing refers to low-budget, under-the-radar niche marketing, using both P2P and any inexpensive top-down methods. Grassroots marketing for large companies may indeed involve a large budget to pay street teams. On the Internet, grassroots marketing includes placing conversational presentations of your product or artist in discussion forums and message boards, placing self-produced videos and music videos online, and encouraging bloggers to write about your artist. This can be accomplished by recruiting fans to be your online street team or e-team.

Table 11.1	Comparison of Grassroots versus Guerilla Marketing	
Grassroots	**Guerilla**
Bottomup/uses people in the market	Low budget
Works on word of mouth	Niche marketing
Avoids mainstream media	Nontraditional
Has street credibility	Avoids costly mainstream media
Used by both small and large companies	Usually used by small companies

Larger corporations have joined the grassroots movement by soliciting media content from customers through contests to "create an ad," "name the product," or "write the slogan." The company Current TV runs contests for companies such as Sony, Toyota, and L'Oreal that encourage customers to submit their homemade 30-second spots. Winners receive a cash award and get their video promoted through mainstream media outlets (Mills, 2007). Large companies also hire members of the target market to spread the word to peers about the artist or product. On the Internet, these online street teams seek out web locations where the target market tends to congregate and infiltrates these areas to introduce the marketing message. *Online street team* members have been known to engage in the following activities to promote recording artists:

1. *Posting on message boards.* Street team members who are actually part of the target market visit relevant sites that allow for message posting and casually encourage readers to "check out my new favorite recording at …." These sites usually have specific rules about the posting of commercial messages and harvesting e-mail addresses (which is universally frowned on by group moderators and system operators). Many groups, however, welcome brief messages from industry insiders notifying interested members of new products. Sometimes street team members do not announce their affiliation with the artist; instead, they engage in communication as if they were just enthusiastic fans. The message postings often are embedded with hot links to the site actually promoting or selling the product. Some user groups require that you register before being allowed to post messages.

FIGURE 11.7
Example of message posting on MySpace.

2. *Visiting and participating in chat rooms.* *Chat rooms* allow for real-time interaction between members, whereas bulletin boards allow individuals to post messages for others to read and respond to. Because these members or users have a mutual interest in the site topic, user groups offer an excellent way for you to locate members of your target market. Marketing professionals as well as web surfers often engage in "lurking" behavior when first introduced to a new users group. *Lurking* involves observing quietly—invisibly watching and reading before actually participating and making yourself known. Often, user groups have their own style and "netiquette" (Internet etiquette), and it's best to learn these rules before jumping in.

3. *Blogging or submitting materials to bloggers.* Successful bloggers (those with a substantial audience) are opinion leaders, and their message can influence the target market. Online street team members will seek out *blogs* that discuss music and reach out to the bloggers to check out their artist and write about the music, much the same way a publicist will seek out album reviews in traditional media. Blogging, however, has the potential for street cred(ibility) that is sometimes absent in mainstream media.

4. *Pitching/promoting to online media.* This is much like pitching to bloggers only more formal. Under these conditions, professional media materials must be supplied to the publication, whether it's an online-only publication or one that also has a traditional media presence. The publication will most likely want a photograph. It's best to supply completed articles and press releases to busy journalists who don't have time to conduct their own research and whose decision to run your article may be influenced by convenience.

5. *Visiting social networking sites and posting materials including music, artwork, and videos.* Major record labels are now using interns and young entry-level employees to maintain their artists' presence on *social networking* sites. These young marketers are responsible for setting up the artist's page on each of these sites, fielding requests from fans to be added to the social group or "friends network," providing updated materials (music, news, photos, videos, etc.), and visiting related pages to engage in street team promotion. Some labels have teamed up with these sites to conduct promotions in the form of contests and giveaways such as encouraging fans to create their own YouTube music video for an artist.

6. *Finding fan-based web sites and asking the site owners to promote the artist.* We address this strategy in the section on fan-based sites. It is often up to online street team members to seek out and identify these fan-based sites. Sometimes they engage in dialogue with the site owner, but often the list of fan-based sites is compiled by the street team member and then passed along to a more senior member of the marketing team. Then these sites are managed by more experienced web marketers who can supply RSS feeds and other technical web assets to the site owner.

7. *Acting as a "clipping service" by scouring the Internet for mentions of the artist.* Google and Yahoo! news alerts have made this task easier, by allowing anyone to set up news alerts based on selected keywords. Any mentions of the artist in third-party online sites should be identified and decisions made about how to incorporate them into the overall web marketing plan. Should you create a link to the article? Should you request permission to reprint? Should you rebut negative publicity?

8. *Finding sites that attract the target market and then working with those sites (see reciprocal links section).* The following section on turning your competition into partners outlines many of the street team tasks performed in this area. The foremost task is finding and identifying where your target market hangs out on the Web—what web sites do they frequently visit? Then you can make marketing decisions about how to work with those sites, whether through advertising on those sites or setting up reciprocal links to generate cross traffic.

9. *Submitting the site to search engines and music directories.* (See the section on search engines.)

10. *Writing reviews of the artist or album on sites that post fan reviews.* Many retail sites such as CDBaby allow customers to post product reviews, including music reviews. It is not uncommon for artists to ask their fans to post reviews on these sites or for a label to have street team members post favorable reviews.

Turn Your Competition into Partners

One way to maximize grassroots marketing is by turning your competition into partners. Because music consumers generally buy several albums per month and are fans of more than one artist, sharing your fan base with similar artists makes sense.

It may be difficult at first locating appropriate partners. The main criterion is that the partners have identical markets of people who would easily cross over from one artist to the next. Cooperative efforts are generally only successful among artists who are all at the same level in their career. A superstar artist's web site is not generally prone to support and feature unknown artists unless the super-star has a personal interest in them. To find artists at the developing level, turn to places like CDBaby and scroll through the artists listed in your artist's genre. For more established artists, Amazon is a good example of a place to find poten-tial co-op partners. On Amazon, it's easy to see the common thread by look-ing at the section titled "customers who bought this item also bought...." Then visit those artists' Amazon page and look at their section on "customers who bought...." After two layers of this, the accuracy of located artists in the same market is somewhat diminished. It is recommended that you only go two layers deep, but that should yield some good prospects to start with, although it will probably yield some overlap. The chart in Figure 11.8 indicates how four con-nections on the artist's Amazon page can ultimately yield 20 potential partners, going only two layers deep.

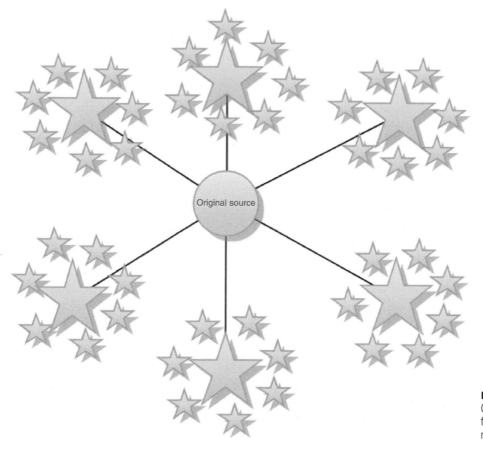

Original source

FIGURE 11.8
Creating a network
for networking and
reciprocal links.

After potential business partners have been identified, the next step is to determine the nature of the partnership and contact each potential partner. The possibilities are described next.

Reciprocal Links

You may just be requesting to trade links, meaning that you will place a link to another's web site on yours, if the potential business partner will reciprocate. This can also be done with banner ads, but that limits the number of partners involved in the promotion. If six people join forces to swap links, visitors to one site will be exposed to the other five artists, thus essentially pooling the customer base. *Reciprocal links*, also known as link exchanges and link swaps, benefit the site in another way. Search engine ranking is partially based on inbound links to your site, indicating popularity and the importance other web sites put on having people visit your site. To find potential link swap partners, visit their web sites and examine their link pages. Determine if a link to your site would be appropriate for them.

In the article "How to Get Reciprocal Links," Allan Gardyne advised the following:

- Find good-quality, complementary sites.
- Place a link to them on your site.
- E-mail the owner of the site a short, friendly note.
- Praise something on their site. If there is nothing to praise, delete them from your list.
- Tell the web site owner you have linked to their site and give them the URL of the page where you've placed their link.
- Ask for a link back to your site, suggesting a page where the link would be appropriate.
- Keep a record of sites you've linked to and requested links from.

Cooperative Efforts, Teaming Up, Promotional Tie-Ins

In addition to link swaps, there are other ways to forge partnerships with former competitors. You can arrange newsletter article exchanges with other newsletter publishers, perhaps reviewing their album in your newsletter and asking for the same consideration or swapping feature articles on each other. Be sure to include some in-context links in the article that will connect back to your web site. If the newsletters are archived online, the impact can be long lasting. Perhaps you want to team up to create a newsletter and then send it to the e-mail lists for both artists. However, it is not advisable to just swap e-mail lists. Your fans have entrusted you with their e-mail addresses and granted you permission to use them to send them information on your artist. If they start receiving unsolicited correspondence from artists they are not familiar with, this could be considered spam, which is certainly unethical and, in some places, illegal.

FIGURE 11.9
Artist co-op: The Nashville Independent Music web page (www.nashvilleim.com).

By pooling efforts, several musicians can split the costs of CD manufacturing, advertising, web site development, and so on. In one instance, a local musician solicited one track each from 10 musicians to create a music sampler. For $300 per artist, the organizer raised $3,000 to cover manufacturing, printing, and marketing costs. Each participant got 100 records for their $300, and the organizer had an instant street team, each member eager to sell as many of their 100 CDs to friends and family as they could. The marketing materials had information on each of the artists and how to purchase more recordings from them and visit their individual web sites.

Artists can also group together to create themed marketing materials, such as a web site that brands a particular style of music and features the artists involved in the co-op. These centralized web sites take the place of the more disordered web rings of the 1990s and afford the co-op the chance to create a brand for the music of all participants by creating a brand name and a consistent image for the co-op. Consumers will quickly learn the meaning of the brand, and the marketing efforts will reach a wider audience.

VIRAL MARKETING

Publicityadvisor.com describes *viral marketing* as a new buzzword for the oldest form of marketing in the world: referral, or word of mouth, which the site has updated to "word of mouse."

Definition and Description

Viral marketing is any strategy that encourages individuals to pass on a marketing message to others, spreading exponentially as one group of people pass on the message to each of their friends. This creates an ever-expanding nexus of Internet users "spreading the word." Viral marketing capitalizes on social networking and the propensity for Internet users to pass along things they find interesting.

Ralph F. Wilson (2000), in his article "Six Simple Principles of Viral Marketing," identified six elements of viral marketing:

1. *Give away something of value.* Giveaways can attract attention. By giving something away up front, the marketer can hope to generate revenue from future transactions. For example, the tag can say, "Win free music at www.yourartist.com."
2. *Provide effortless transfer to others.* According to Wilson, "The medium that carries your marketing message must be easy to transfer and replicate."
3. *Expand exponentially.* Scalability must be built in so that one message transmitted to 10 people gets passed on to 10 each, for 100 new messages.
4. *Exploit common motivations and behaviors.* Success relies on the basic urge to communicate and share experiences and knowledge with others.
5. *Utilize existing communication networks.* Learn to place viral messages into existing communications and the message will "rapidly multiply in its dispersion."
6. *Take advantage of others' resources.* A news release reprinted elsewhere will include the viral message and perhaps the link.

Examples of Effective Viral Campaigns

IncrediMail, designers of e-mail software, worked with MGM to create an online viral buzz for the release of the new Rocky movie by giving away various "Rocky Balboa"–themed skins for toolbars, e-mail backgrounds, wallpapers, and screensavers.

The viral marketing firm Fanscape was contracted to promote Clear Channel's new music web site New! to unsigned bands to encourage them to upload their music to the site. The incentive was the opportunity for the band to expand its fan base through the site and increase its chance of being noticed by radio station program directors, because the music was to be made available publicly for streaming. The "most played" songs are prominently featured. Fanscape "reached out to unsigned bands by targeting them on MySpace, Facebook, Garage Band, etc. to personally invite them to check out New! and submit their music to the site, and to possibly be heard by radio program directors throughout the country" (www.womma. com). Fanscape generated more than 1,500 music submissions in two weeks.

Ways to Incorporate Viral Marketing into Your Communication Campaign

The five most commonly used viral marketing methods are (1) the e-mail signature, (2) screensaver giveaways, (3) the tell-a-friend script, (4) the use of message boards, and (5) writing articles and allowing reprints.

- *E-mail.* The most common form of viral marketing is through e-mail signature attachments. Hotmail.com and Yahoo! have successfully employed this technique by appending their message to the bottom of every e-mail generated by its users. Hotmail offers free e-mail service to users but also places a viral tag at the end of each message that its users generate. If that message is passed along, the advertising tag goes with it.
- *Customized screen saver.* The web site www.2createawebsite.com suggests that an attractive screensaver that is distributed for free and is imbedded with links to your web site will promote your products while offering something of interest to the user. For example, photos of a band in concert could be developed into a screensaver and include links to the artist's web site. The screensaver is then offered for free to fans through the e-mail list and on the web site. There is always the possibility that fans will pass the screensaver program along to their friends. Other giveaway viral tools include skins for toolbars, e-mail backgrounds, and wallpapers.
- *Tell-a-friend script.* These are usually found on the artist's web site and can encourage visitors to pass along your information to their friends. Jeanne Jennings of the ClickZ Network suggests several features to increase the success rate of the "tell-a-friend function." These can be found in Chapter 7. She suggests making it prominent, making it visual, including address book functionality, adding privacy and copyright concerns, offering senders their own copy, supporting personalization, and promoting your business goals (www.2createawebsite.com/traffic/viral-marketing.html).

Yahoo! Music "Get Your Freak On"

The growing popularity of online video-sharing and digital music created a perfect combination for Yahoo! Music to engage music enthusiasts with a one-of-a-kind promotion. Yahoo! Music created a program called "Get Your Freak On" where fans are invited to star in user-generated music videos for their favorite artists and songs.

This ongoing program has resulted in wildly popular videos for music stars like Jessica Simpson, Christina Aguilera, and 'Lil Jon, but it's the video for megastar Shakira that put this program on the map. Fans of the Colombian-born recording artist were asked to submit short, hip-shaking video clips to be included in a fans-only music video for the song "Hips Don't Lie."

After thousands of fans submitted their videos, Yahoo! and Epic took the best entries and created a video mashup that was exclusively available on Yahoo! Music. The fans-only video quickly raced to the number-one spot on Yahoo! Music, and even surpassed the popularity of the song's actual video. The user-generated music video was viewed more than one million times in the first few weeks of release, and has been seen more than 12 million times to date. At the time of the premiere of the fan-video, Shakira's actual video was #1 on Yahoo! Music, and the following week her record jumped 92 spots to become the 6th best selling album.

Yahoo! recognized the potential for this high-profile promotion to engage Yahoo! Music loyalists and spread virally throughout the Yahoo! network. The marketing team created a comprehensive campaign to kick-start the WOM nature of this promotion, including banner ads running across Yahoo, integration on Yahoo.com, and editorial placement in Yahoo! Music and within Yahoo!'s Buzz module.

Yahoo! also played close attention to Shakira's fan base and integrated the campaign on "Yahoo! En Espanol" and worked with Shakira Fan Clubs.

Credit Information

Client: Yahoo!
Agency: N/A
Budget: $15,000
Date of Campaign: Undisclosed

FIGURE 11.10
Yahoo! Video viral marketing campaign (permission from Yahoo!).

■ *Use of message boards.* Jeanne Jennings suggests that you participate in as many relevant forums, chat rooms, and *message boards* as possible. The more messages posted about your artist and the more often their name is associated with messages of value, the more likely people will seek out and visit the artist's web site. Some forums will allow a link in the signature file. For others that do not allow the use of links or the overt promotion of products, be sure to pick out a user name that can be easily traced to the web site. For example, the Sauce Boss would use that signature on all message board postings, expecting that interested readers would be able to easily find his web site at www.sauceboss.com.

- *Write articles of interest and allow reprints.* Because webmasters are always looking for good content, a well-written article on a topic of interest will be enthusiastically received and published. Links attached to the author name, at various points within the article, and in references at the end will create traffic to the web site. As the article is passed around electronically, anyone interested in the article will be exposed to the links. To find viral marketing articles, visit www.wilsonweb.com/cat/cat.cfm?page=1& subcat=mm_viral.

FAN-BASED WEB SITES

Fan-based sites are unofficial web sites or web pages featuring an artist. Often they can be an extension of the fan's personal web site that also features other aspects of the fan's life. But many times they are dedicated fan sites honoring the creator's favorite artist and fostering a community of fans coming together. This ready-made street team represents some of the most motivated fans who can help propel an artist's career.

Importance of Fan-Based Sites

Often, the value of fan-based sites is overlooked in the grassroots marketing plan. They are obviously more common for well-known artists, but they exist for lesser-known artists in particular subcultures. At first, record labels were unsure how to deal with these sites, many of which had outdated or incorrect information about the artist. But labels quickly learned the value of this self-proclaimed street team—these are motivated fans who want to spread the word. Labels and webmasters now actively provide information to these sites in the form of updated photos, press releases, and RSS feeds (see the section that follows) so that they become part of the artist's online network.

For lesser-known artists, it might be necessary for the artist and management team to request that active fans create these sites. Some record labels have been known to have their interns set up and manage these sites or to create fan-based pages on social networks that appear to have no connection to the sponsoring label.

How to Find and Encourage Fan-Based Sites

To encourage and maximize fan-based web sites, marketers for established artists should scour the Internet using search engines to find these fan-based pages. Then, contact the web site owner of each site and ask them to join the "online network" to receive updates. Most fans will jump at the chance to receive official information and photos from the artist's management team before the general public has access to them.

Figure 11.11 presents an example of an R.E.M fan-based site, which shows how fans can successfully create and promote a site supporting their favorite artists. Care must be taken that no copyright infringement takes place and that all materials posted on the fan site are accurate and current. Notice how this site does not use officially licensed logos typically found on the official site, and no attempt has been made to deceive the visitor into thinking this might be the official artist site. Figure 11.11.

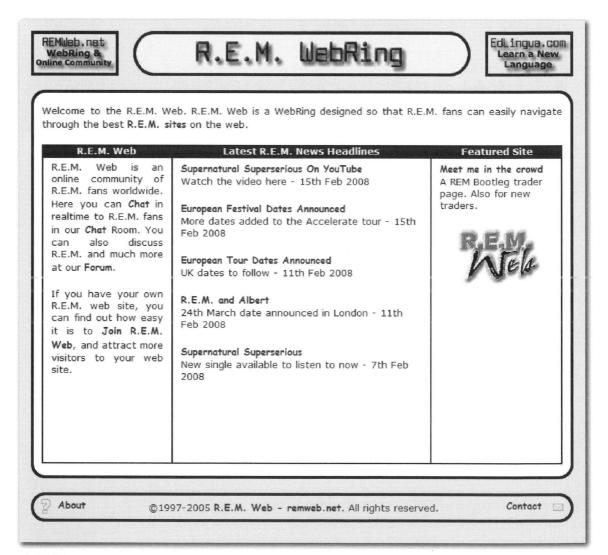

FIGURE 11.11
Example of a fan-based site. (Reprinted by permission.)

Tips on Helping Fans Maintain Their Fan-Based Sites

Social networking sites have recently facilitated fan-based music sites for users who may not have the savvy or initiative to create a standalone web site honoring their favorite music. Buzznet.com is one such social networking site that encourages its members to support and promote music on the site and reward active trendsetters through the Buzzmob community. The Buzznet web site states, "The Buzzmob is the e-street team of Buzznet's most loyal fans who help promote Buzznet, bands and e-scene personalities across the Net. The Buzzmob is a place for Buzznet fanatics to get to know one another and help Buzznet grow" (www.buzznet.com/www/buzzmob/?t=footer|buzzmob).

RSS FEEDS TO FAN-BASED SITES

An *RSS feed* (really simple syndication) is frequently updated syndicated content published by a web site that appears on other web sites. It can also be used for distributing other types of digital content such as pictures, audio, or video. On MySpace, when a member changes out his or her primary photo, the updated photo appears in all the member's friends' "my friend spaces" and "bulletin spaces." Once information is in the RSS format, an RSS-aware program can check the feed for changes and react to the changes in the appropriate way.

Webmasters for recording artists have found that RSS feeds are an efficient way to update information on many "satellite" fan sites with little effort. This effort keeps the information on the artist current and accurate.

ADVERTISING

To generate traffic to a web site, it might be necessary to advertise on the Internet. Your return on investment (ROI) will help determine whether advertising is appropriate or not. For an emerging artist, most of your web site hits will come from word of mouth—announcing at gigs and posting your web address on all materials. Advertising would be more appropriate for a concert promoter or a record label, especially one that specializes in a niche form of music, such as Celtic. For example, when you type "Celtic music" into the Google search box, www.celticmusicusa.com pops up as a sponsored link. By visiting that web site, you learn of the vast catalog of music offered through this site. Because most advertisers pay based on the number of people who click on their ad, thus sending them to the advertiser's site (see the discussion on cost-per-click, presented later), those costs must be weighed against potential sales as a result of the advertising program. If, for example, only one in a hundred visitors to your site—generated through advertising—makes a purchase, and the average purchase is $10, how much would you be spending in advertising for each sale? If you pay 10 cents for each click, you have spent $10 in advertising for each sale—hardly worth the money. Sometimes companies will advertise anyway in hopes of building traffic, generating some repeat visits to the site, creating online word of mouth, and raising the site's profile in future search engine results. See Figure 11.12.

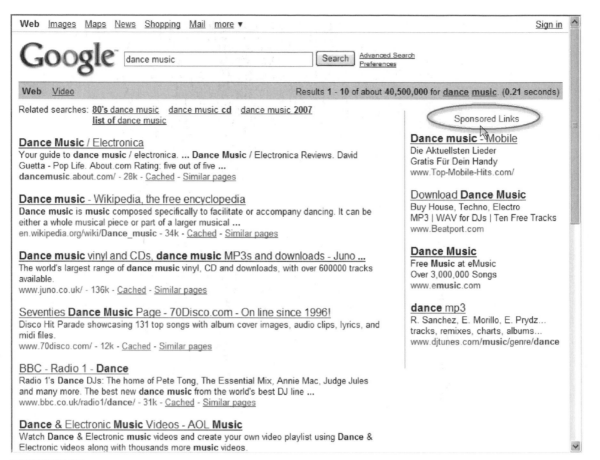

FIGURE 11.12
Example of keyword-driven ads on Google.

Most Internet advertising takes the form of sponsored links, replacing the earlier style of choice—banner ads. Banner ads are basically a graphical advertisement placed across the top or down the side of a web page (also called a sidebar ad) that is linked to the advertiser's web site. Marshall Brain, in his article "How Web Advertising Works," stated that "In the beginning [of the web] advertising on the Internet meant banner ads." This fueled a financial boom for the dot-com industry as advertisers switched more of their budgets from traditional media to the Internet. After this experimental phase, advertisers began to realize that banner ads were not as effective as magazines or "spot" commercials and scaled back on web advertising (Brain, 2002).

Advertisements would appear on popular content sites to pitch products available on e-commerce sites. Originally, these ad rates were set based on how it is done in the magazine industry—that is, they were based on the number of impressions. Thus, the advertiser paid based on the number of people who saw the ad. Now, the most popular form of charging for such advertising is "cost per click" (CPC), also known as pay per click (PPC). With CPC, the advertiser is charged a small fee each time a potential customer clicks on the ad and is taken to the advertiser's web site. This has proved more effective because the advertiser is paying only for those potential customers who respond to the ad by clicking on it and being directed to the advertiser's site. Google has popularized the use of sponsored links and offers web sites with similar content the ability to feature Google ads from sponsors who sign up for Google's AdWords.

Here is how Google AdWords and Yahoo! Marketing Solutions work. As stated earlier, advertisers pay on a per-click basis; in other words, they pay a few cents for each time a web visitor clicks on their sponsored link. The advertiser enters in a series of *keywords* (search terms or words their customers are likely to use in a search engine when looking for a particular product or type of web site). Advertisers place bids to have their ad strategically located in the sponsored links category on search engine results. The advertiser can actually create a list of terms and bid independently on each one. The highest bid for that particular set of search terms has the top spot. You may not want to be the top spot for blues music if you depend on live shows for income because blues fans from all over would be likely to click to your site only to learn that you are not performing in their area. But if the term "blues music" was combined with "East Texas" and you are performing in that area, then perhaps you want one of the top advertiser spots. Finding the right keywords and combination of keywords may take a bit of trial and error at first.

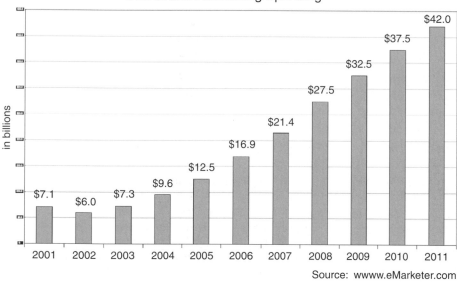

U.S. Online Advertising Spending

in billions

$7.1 (2001) $6.0 (2002) $7.3 (2003) $9.6 (2004) $12.5 (2005) $16.9 (2006) $21.4 (2007) $27.5 (2008) $32.5 (2009) $37.5 (2010) $42.0 (2011)

Source: wwww.eMarketer.com

FIGURE 11.13
U.S. online ad spending. (Source: eMarketer.com.)

Contextual Advertising

Contextual advertising is defined as advertising on a web site that is targeted to the specific individual who is visiting the web site based on the subject matter of the site and then featuring products that relate to that subject matter. For example, if the user is viewing a site about playing music and the site uses contextual advertising, the user might see ads for music-related companies such as music stores. Google has added AdSense as a way for web site owners to feature relevant advertising on their sites and share in the CPC revenue generated by sponsors. The source of these ads comes from the AdWords program, so that those who sign up for AdWords can specify if they want their ad to appear on these related web sites—the *content network*. Google's web site states, "A content network page might be a web site that discusses a product you sell, or a blog or news article on a topic related to your business."

US Online Advertising Spending by Format 2006-2011
(percent of total online advertising spending in $ billions)

	2006	2007	2008	2009	2010	2011
Paid search	40.3%	40.3%	40.0%	39.8%	39.8%	39.5%
Display ads	21.8%	21.9%	21.5%	20.5%	20.0%	19.5%
Classified	18.1%	17.0%	17.0%	16.9%	16.8%	16.5%
Rich media/video	7.1%	8.2%	9.5%	11.0%	11.9%	13.1%
Lead generation*	7.8%	8.2%	8.3%	8.6%	8.8%	8.8%
E-mail	2.0%	2.0%	1.8%	1.7%	1.6%	1.5%
Sponsorships	2.9%	2.5%	2.0%	1.5%	1.3%	1.2%
TOTAL	$16.9	$21.4	$27.5	$32.5	$37.5	$42.0

*Using marketing tools and/or advertising techniques in order to acquire leads.

FIGURE 11.14
U.S. online advertising spending by format. (Source: www.emarketer.com.)

Table 11.2	U.S. Online Advertising Spending by Format, 2006–2011*					
	2006	**2007**	**2008**	**2009**	**2010**	**2011**
Paid search	40.3%	40.3%	40.0%	39.8%	39.8%	39.5%
Display ads	21.8%	21.9%	21.5%	20.5%	20.0%	19.5%
Classified	18.1%	17.0%	17.0%	16.9%	16.8%	16.5%
Rich media/video	7.1%	8.2%	9.5%	11.0%	11.9%	13.1%
Lead generation*	7.8%	8.2%	8.3%	8.6%	8.8%	8.8%
E-mail	2.0%	2.0%	1.8%	1.7%	1.6%	1.5%
Sponsorships	2.9%	2.5%	2.0%	1.5%	1.3%	1.2%
TOTAL	$16.9	$21.4	$27.5	$32.5	$37.5	$42.0

*Percentage of total online advertising spending in billions of dollars.
Source: www.emarketer.com.

GLOSSARY

AdSense – An advertising program by which Google acts as the intermediary between content sites and web advertisers.

BCC (blind carbon copy) – A way to send an e-mail message without revealing the recipient's address.

Blogging or blog – Short for weblog, it is a personal online journal that is frequently updated and intended for general public consumption.

Chat room – Part of a web site that provides a venue for communities of users with a common interest to communicate to the group in real time.

Click through rate (CTR) – The ratio/percentage of the number of times an ad is clicked divided by the number of times an ad is viewed.

Clipping service – (see Press clippings) A service that compiles relevant press clippings for a client.

Contextual advertising – Advertising on a web site that is targeted to the specific individual who is visiting the web site.

Cost per click (CPC) – The amount of money you pay to a search engine for one click on your ad.

Electronic press kit (EPK) – A press kit equivalent in electronic form.

E-zines – An electronic magazine, whether posted via a web site or sent as an e-mail newsletter. Short for electronic magazine or fanzine, some are electronic versions of existing print magazines, whereas others exist only in the digital format.

Fan-based web sites – Unofficial web sites or web pages featuring an artist, often an extension of the fan's personal web site.

Grassroots marketing – Unconventional street-level marketing using word-of-mouth influence and that of opinion leaders to disseminate a marketing message among potential customers.

Guerilla marketing – Low budget, under-the-radar niche marketing, using both peer-to-peer (P2P) and any inexpensive top-down methods.

Keyword/keyword phrase – The term or phrase that a user types in the search box of a search engine to receive more information that relates to that term/phrase.

Lurking – Invisibly hanging out in chat rooms instead of actively participating in the discussion.

Mail merge – The process of producing a personalized e-mail letter for each person on a mailing list by combining a database of names and addresses with a form letter.

Message boards (online) – A script on a web site with a submission form that allows visitors to post messages (called "threads" or "posts") on a web site for others to read. These messages are usually sorted within discussion categories, or topics, chosen by either the host or the visitor.

Online street team – Similar to a record label street team, but its members engage in online word of mouth and other Internet marketing campaigns.

Peer to peer (P2P) – A type of transient Internet network that allows a group of computer users with the same networking program to connect with each other and directly share files from one another's hard drives.

Podcast – A digital media file that is distributed over the Internet for playback on portable media players and personal computers. Unlike real-time radio broadcasting, the podcast is usually played back at the convenience of the listener.

Press clippings – Relevant excerpts cut from a newspaper or magazine.

Reciprocal link – A mutual link between two objects, commonly between two web sites, in order to ensure mutual traffic.

Reprint – A separately printed article that originally appeared in a larger or previous publication.

RSS feed (really simple syndication) – A frequently updated and automated syndicated content feed provided by one web site to others for distributing digital content such as text, pictures, audio, and video.

Search engines – Online directories of web sites and web pages that web visitors use to find a topic of interest or specific site.

Search engine optimization (SEO) – A term that defines the sum of activities you run in order to promote your web site as high as possible among the organic results on a search results page.

Simulcast – Simulcast is a contraction of "simultaneous broadcast" and refers to programs or events broadcast across more than one medium.

Social networking – The practice of interacting with and expanding the number of one's business or social contacts by making connections typically through social networking web sites such as MySpace and Facebook.

SoundExchange – An independent, nonprofit performance rights organization that is designated by the U.S. Copyright Office to collect and distribute digital performance royalties for featured recording artists and sound recording copyright owners when their sound recordings are performed on digital cable and satellite television music, the Internet, and satellite radio.

Viral marketing – A marketing phenomenon that uses online social networking to facilitate and encourage people to pass along a marketing message voluntarily and exponentially. Sometimes the marketing message is imbedded in, or attached to, a message that participants find interesting and willing to forward to others.

Webcasting – The concept of broadcasting much like a radio station, but via the Internet in lieu of a wireless signal.

REFERENCES AND FURTHER READING

Atkinson, Robert. (2007, July 12). The day the (web) music died, http://www.huffingtonpost.com/robert-d-atkinson-phd/the-day-the-web-music-d_b_56029.html.

Borg, Bobby. (2005). Web radio stations and getting played: A viable medium for new bands, www.indie-music.com/modules.php?name=News&file=article&sid=3604.

Brain, Marshall. "How web advertising works." April 08, 2002, http://computer.howstuffworks.comweb-advertising.htm. (June 21, 2007).

Brooks, Rich. 7 Reasons to start an Email newsletter today. Flyte New Media, www.flyte.biz/resources/articles/0303.php.

Dawes, John, and Sweeney, Tim. (2000). The Complete Guide to Internet Promotion for Artists, Musicians and Songwriters. Tim Sweeney and Associates, Temecula, CA.

Deitz, Corey. (2005). Jimmy Buffet's Radio Margaritaville moves to SIRIUS Satellite Radio, http://radio.about.com/od/siriussatelliteprograms/qt/bljimmybuffet.htm.

Email marketing software.org, www.emailmarketingsoftware.org/learn/email-marketing-software.html.

Farrish, Bryan. Radio Airplay 101: Traditional radio vs. the Web, www.radio-media.com/song-album/articles/airplay14.html.

Free Web Hosts.com. www.freewebhosts.com.

Foster, Jane. Google AdWords 101, www.article99.com

Gardyne, Allan. (2005, December 20). How to get reciprocal links, http://www.associateprograms.com/articles/48/1/How-to-get-reciprocal-links.

G-Lock EasyMail. Tips for successful Email marketing campaign, www.glocksoft.com.

Harbin, Jennifer. (2007). Universal Music Group new media intern guide. Unpublished manuscript.

Hitwise. (2007, March 14). IMeem and Bebo are the fastest movers, www.hitwise.com/press-center/hitwiseHS2004/socialnetworkingmarch07.php.

Jennings, Jeanne. (2007, June 4). Optimizing the "e-mail this" marketing opportunity. The ClickZ Network. http://www.clickz.com/showPage.html?page=3626029.

Mills, Elinor. (2007). Advertisers look to grassroots marketing, news.com.com/2100-1024_3-6057300.html.

Monash, Curt. (2004). Search engine ranking algorithms. *The Spider's Apprentice.* www.monash.com/search_engine_ranking.html.

Nevue, David. (2005). How to Promote Your Music Successfully on the Internet. Midnight Rain Productions, Eugene, OR.

Search Web Services.com.

Sullivan, Danny. (2007, March 14). Search engine placement tips, http://searchenginewatch.com/showPage.html?page=2168021#position.

Sullivan, Danny. (2007, March 15). How search engines rank web pages. http://searchenginewatch.com/showPage.html?page=2167961.

Taglieri, John. (2007, June 6). Internet radio—the new age of airplay. http://www.musesmuse.com/musicformats-4.html.

Walker, Chris. (2006). Search engine optimization versus pay-per-click, www.article99.com.

Wilson, R. (2000, February 1). The six simple principles of viral marketing. *Web Marketing Today,* p. 70.

Wilson, Ralph. (2000, July 1). Using banner ads to promote your website. *Web Marketing Today.* http://www.wilsonweb.com/articles/bannerad.htm.

www.adwordshowtos.com/learn/google-adwords-glossary.html.

ZDnet.com. (2008, February 20). 4.85 billion hours spent on online music in 2007. http://blogs.zdnet.com/ITFacts/?p=13961#.

CHAPTER 12
Social Networking Sites

Social networking sites are defined as "web sites that allow members to construct a public or semipublic profile and formally articulate their relationship to other users in a way that is visible to anyone who can access their file" (Boyd and Ellison, 2007). Social networking sites allow anyone to have a personal web presence. Characteristics that have made social networking sites so popular include the following:

1. They allow for self-expression.
2. They require little or no knowledge of web design.
3. They allow for social interaction and networking.
4. They are free or inexpensive to use.

RISE AND POPULARITY OF SOCIAL NETWORKING SITES

The first significant social networking site was classmates.com, founded in 1995. Classmates.com allows for high school and college classmates and graduates to stay in touch. The site is still going strong today but has been joined by, among others, Friendster.com, founded in 2002, MySpace, founded in 2003, and Facebook, founded in 2004. According to Hitwise, a company that specializes in online web traffic and marketing, visits to social networking sites accounted for 6.5% of all Internet visits in February 2007. At that point in time, MySpace accounted for 80% of the market share, with Facebook holding 10% of the market. Other services such as Bebo, BlackPlanet, Classmates, Friendster, Orkut, and imeem accounted for less than 1% each. The same press release stated that "Buzznet and iMeem are succeeding in building communities around music" (www.hitwise.com/press-center/hitwiseHS2004/socialnetworkingmarch07.php). By April 2007, Facebook had grown its market share to 11.5%, doubling its traffic since opening up its service to Internet users without school affiliation. The year 2007 was a watershed for music-oriented social networks as a host of upstarts were all vying to be the next MySpace. By the end of 2007, the other social networking services had begun to encroach on MySpace's commanding lead, which had declined slightly to 76% for the year and 72% for December 2007. Facebook had continued its growth to 16% by December, with Bebo and BlackPlanet moving over the 1% mark. In early 2008, imeem acquired fledgling online retailer SNOCAP.

The article "Could Social Networking Save the Music Industry," on www.cio-today.com (January 2007), professed that the new generation of social networking sites that emphasize music might be just what the industry needs to pull teens and young adults away from illegal file-sharing networks and back into the world of legal music consumers. These advertiser-supported sites offer what cio-today calls "a better form of free" for consumers.

MYSPACE

Because MySpace has the lion's share of social networking traffic at the time of this writing, it should be looked on as the most important social networking site worthy of marketing efforts. However, this could change as other sites vie to compete in the marketplace, hoping to offer new and exciting services to lure web visitors away from MySpace and Facebook.

Signing Up and Setting Up Your Band Site

For artists, MySpace's setup is ideal for promoting music. Some of the features for artists include a built-in music player, user ratings, reviews, artist rankings, featured bands, show listings by location, and music videos. MySpace launched its own record label in 2005 with arrangements made for distribution through Universal Music Group. Through the online retailer SNOCAP, MySpace now offers all its independent artists the chance to sell downloads from their MySpace page.

To get started, you will want to use your e-mail address to sign up for an artist account and not a personal account. If you already have a personal account, you will need to use a different e-mail address to sign up for the artist account. If you wish to keep your current e-mail address the same, you can delete your personal account on MySpace or change the e-mail address on your personal account. To sign up for an artist site on MySpace, locate and click on the "music" link on the menu bar (not the search menu in the top banner). When the MySpace music page opens, click on "artist signup." You will get a chance to specify up to three genres under which your artist will be categorized, but on the signup page, you pick the primary genre and enter the band's name. This will become the artist's address at www.myspace.com/artistname. To complete the registration, you get the chance to list the artist's web site address, the three genres under which you want the artist categorized, and the artist's label affiliation.

FIGURE 12.1
MySpace page for jazz
artist Inga Swearingen.
(Printed with permission
from MySpace.)

Inga Swearingen's new album **"Reverie"** with the Bill Peterson Trio now available at www.cdbaby.com

Inga Swearingen's debut album **"Learning How To Fly"** with Trio 14 also available at www.cdbaby.com

You have the option of including crucial artist details in the artist profile, such as a bio, list of band members, upcoming shows, musical influences and comparisons, and the URL of the band's official web site. At this point, you get the chance to invite friends to sign up and to upload photos and audio files, but you must use the MP3 format for audio files. Use the "manage songs" page to upload up to six songs. You can specify whether listeners are allowed to

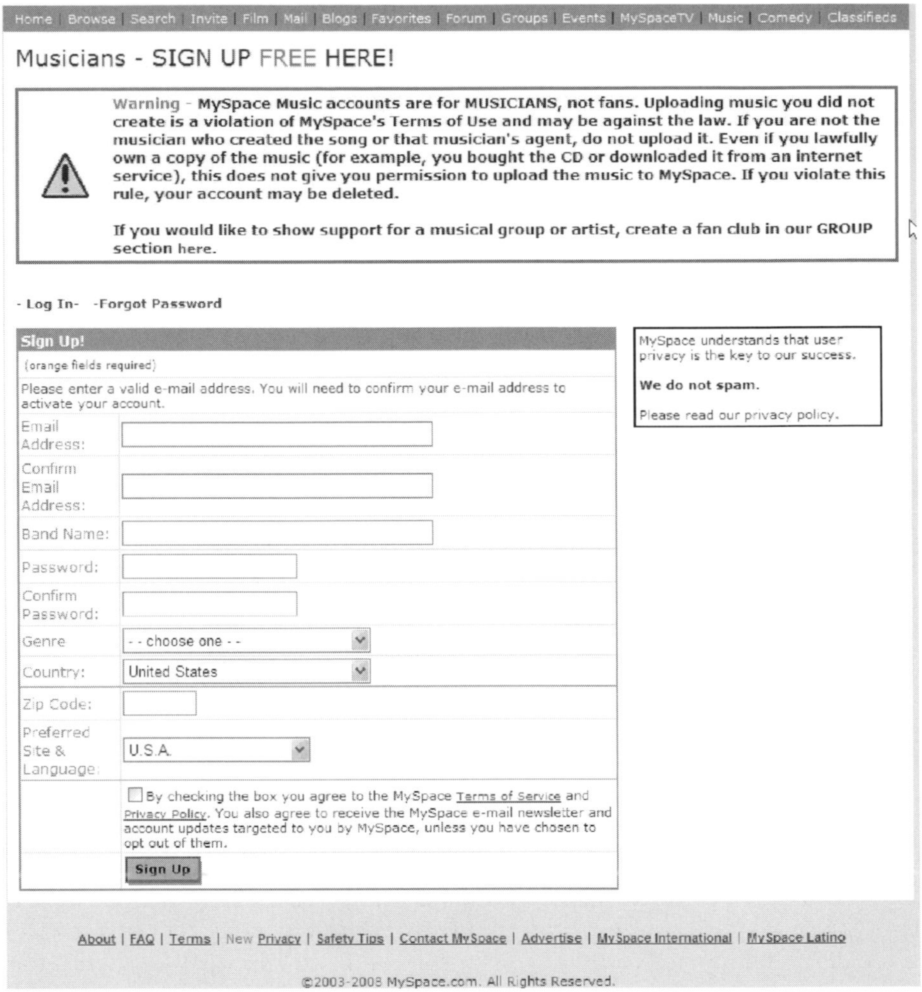
Home | Browse | Search | Invite | Film | Mail | Blogs | Favorites | Forum | Groups | Events | MySpaceTV | Music | Comedy | Classifieds
Musicians - SIGN UP FREE HERE!
Warning - MySpace Music accounts are for MUSICIANS, not fans. Uploading music you did not create is a violation of MySpace's Terms of Use and may be against the law. If you are not the musician who created the song or that musician's agent, do not upload it. Even if you lawfully own a copy of the music (for example, you bought the CD or downloaded it from an internet service), this does not give you permission to upload the music to MySpace. If you violate this rule, your account may be deleted.
If you would like to show support for a musical group or artist, create a fan club in our GROUP section here.
- Log In- -Forgot Password
Sign Up!
(orange fields required)
Please enter a valid e-mail address. You will need to confirm your e-mail address to activate your account.
Email Address:
Confirm Email Address:
Band Name:
Password:
Confirm Password:
Genre
- - choose one - -
Country:
United States
Zip Code:
Preferred Site & Language
U.S.A.
By checking the box you agree to the MySpace Terms of Service and Privacy Policy. You also agree to receive the MySpace e-mail newsletter and account updates targeted to you by MySpace, unless you have chosen to opt out of them.
Sign Up
MySpace understands that user privacy is the key to our success.
We do not spam.
Please read our privacy policy.
About | FAQ | Terms | New Privacy | Safety Tips | Contact MySpace | Advertise | MySpace International | MySpace Latino
©2003-2008 MySpace.com. All Rights Reserved.

FIGURE 12.2
MySpace artist sign-up registration.

download the song or merely stream it to their computers. You can also assign an image to each of the songs in your profile. For future editing of songs, band details, tour dates, and so on, click on the "edit profile" link. When adding a song to the site, you have the option to allow users to add *your* song to *their* profile, to auto-play the first song when people visit your page, or to randomize song play. Depending on the parameters you set, your visitors can have the option to listen to the songs in your music player, rate your songs, comment on them, download a copy, view the lyrics, or click a link to add the song to their own MySpace page.

To change the default photo on your site, click on the "add/edit photos" link. Select the photo from the correct album by double clicking the photo, and then click on "set as default." Through the "manage address book link" you have the

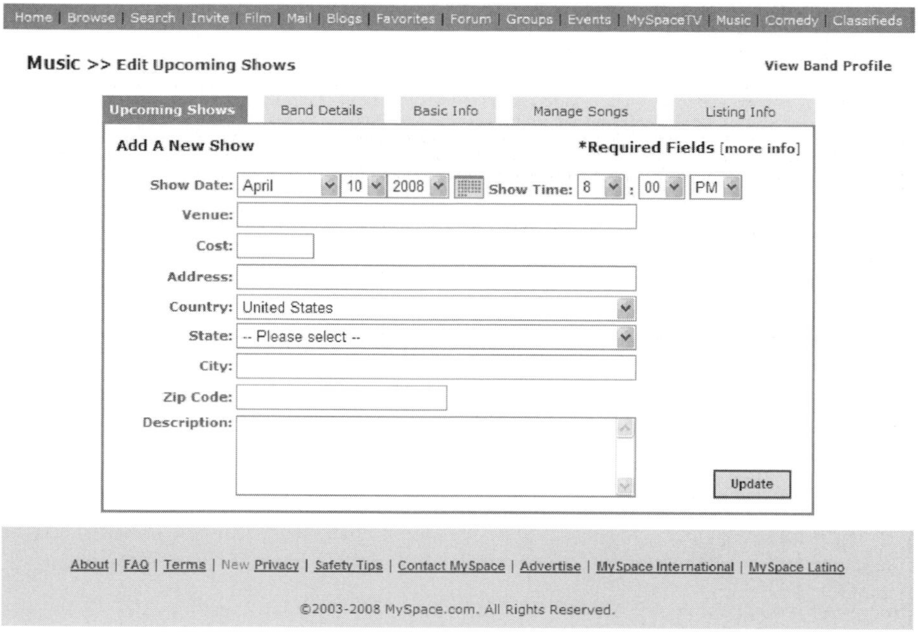

FIGURE 12.3
Edit upcoming shows
or add a show date
on MySpace. (Printed
with permission from
MySpace.)

opportunity to build an address book of MySpace users and manage outgoing and incoming mail. To collect e-mail addresses of visitors, you can embed a mailing list signup form into the profile page. MySpace offers much more than internal networking. The simple web address assigned to each artist makes it possible to use the MySpace page as the main Internet presence for an artist, allowing for music sampling, posting of the tour schedule, blogging, and much more. Or for artists who have a web site but need help setting up a listening page or an online retail presence, they can now simply link to their MySpace page for this service. It is advised that if the MySpace page is used primarily for music sales, keep the clutter on that page to a minimum so visitors can focus on the music.

Formatting the artist's page can be a challenge. MySpace offers flexibility with page formatting, but within parameters. The default profile for artists features the artist name and photograph on the left side of the page, with a music player on the right side. Underneath the photo on the left side are the communications buttons and general information. Beneath the music player on the right side is the SNOCAP store. Customizing the page, also known as "pimping," can involve simple changes in scheme color, and font style, or it can involve drastic template makeovers, many of which are available either free or for a small fee on the Internet. Make sure that your new layout follows the design rules presented in Chapter 5. Too many MySpace sites have been loaded down with background graphics that mask portions of the text and render it unreadable. Installation is usually simple: Pimp-My-Profile.com states, "To install a layout, copy & paste the layout code into the 'about me' section of your profile" (www.pimp-my-profile.com).

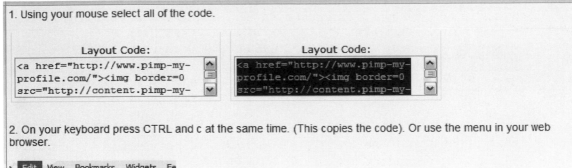

1. Using your mouse select all of the code.

2. On your keyboard press CTRL and c at the same time. (This copies the code). Or use the menu in your web browser.

3. Go to the place where you want to paste the code (this example shows myspace). Put your cursor in the area where you want the code to go.

4. Press CTRL and V on your computer at the same time, or use the controls in your browser.

FIGURE 12.4
Installing code from a third-party developer to a MySpace page. (Printed with permission from MySpace.)

Subtle customization can be achieved with a few basic HTML commands (see Chapter 6). Use the "Edit profile" feature to access fields such as the "bio" section under band details. Basic HTML language can be inserted in the text to change fonts and text styles and to include links. Then click preview to see the changes. Not all fields read and respond to all HTML commands, so some experimentation is necessary to see what works and looks good. You can also use an outside HTML editor, where you can set the text in a WYSIWYG program and then copy and paste the resulting code. Images can be added to many sections of a MySpace page. Keep photos at the proper size, no greater than 450×550 pixels. Also, thumbnails of the profile photo will appear on friends' MySpace pages, so don't use an image with lots of small detail. Keep the composition of the profile photo to an image that would resemble what is found on a postage stamp.

Because MySpace opened up its platform to third-party developers, many widgets are available to spice up a web site. Again, integrating these into the MySpace page is as simple as copying a prewritten code into the proper place or clicking on an icon on the third party's site (see widgets, Chapter 6).

FIGURE 12.5
SNOCAP on a MySpace music page. (Printed with permission from MySpace.)

To sell music on MySpace through SNOCAP, you must first set up a SNOCAP account (at the time of this writing, the $30 fee was waived for the first year) and upload your music. Once that is completed, a MySpace Music Store will be automatically created on your artist's MySpace page directly under the MySpace music player. The agreement with SNOCAP is nonexclusive, meaning that you are still free to sell the same music through other outlets. Your fans will be allowed to purchase music directly from your profile page at a price you specify. The system allows you to offer up to 100 tracks with 30-second samples of each. More information on SNOCAP can be found at www.snocap.com/about/faq.

Networking and Promotional Techniques

Networking on MySpace is accomplished by requesting to add a "friend." A friend is someone you exchange profiles with, so that your profile and link appear on your friend's page and your friend's profile and link appear on your page (Nevue, 2005). "Theoretically speaking, the more 'friends' you have, the larger 'network' you have to promote your music to (Nevue, p. 109). You can build your network by going to friends' pages or pages of artists similar to your own and

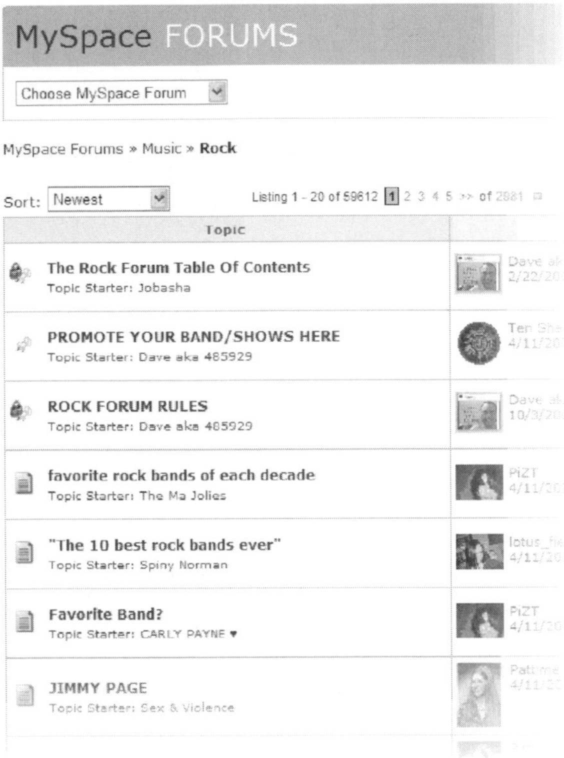

MySpace FORUMS

Choose MySpace Forum ▾

MySpace Forums » Music » **Rock**

Sort: Newest ▾   Listing 1 - 20 of 59612 **1** 2 3 4 5 >> of 2981

Topic	
The Rock Forum Table Of Contents Topic Starter: Jobasha	Dave ak 2/22/20
PROMOTE YOUR BAND/SHOWS HERE Topic Starter: Dave aka 485929	Ten Sh 4/11/20
ROCK FORUM RULES Topic Starter: Dave aka 485929	Dave ak 10/3/20
favorite rock bands of each decade Topic Starter: The Ma Jolies	PiZT 4/11/20
"The 10 best rock bands ever" Topic Starter: Spiny Norman	lotus_f 4/11/20
Favorite Band? Topic Starter: CARLY PAYNE ♥	PiZT 4/11/20
JIMMY PAGE Topic Starter: Sex & Violence	Pattime 4/11/2

FIGURE 12.6
Example music forums.
(Printed with permission
from MySpace.)

see who is in that person's Friend Space. Then you can click on the photo to be directed to that person's profile and page. If any of these new contacts seem like good potential fans, you can request to add them to your artist's Friend Space (Brothers and Layton, 2007). If they are unfamiliar with your artist, they may visit your artist's page and check it out before deciding whether to approve the friend's request. (More information can be found in Chapter 11.)

It is also possible to find members of the target market by using the music forums or the directory to find similar artists. Then by visiting their sites, you can again look through their Friend Space.

In the book *MySpace Music Marketing*, author Bob Baker suggested the following to locate your potential fans:

- Use the search feature to input specific keywords related to your artist, their sound, or other things of interest to their potential fans.
- Look through the Friends list of similar artists to find potential fans. By clicking on "view all of artist X's friends," you can scroll through and click on the photos to get their individual profiles.
- Look through the comments section of similar artists for particular comments that may indicate whether or not this person is a potential fan.
- Use the browse feature to search for potential fans based on demographic characteristics.

Once you have developed a list of friends, cultivate these new fans by engaging in many of the opportunities MySpace has to offer. One is that any MySpace user can add a streaming song to his or her personal profile page. For your artist's music that has been uploaded to the MySpace music page, be sure to allow users to add your song to their personal page for streaming. Then, whenever anyone visits that person's page, the song will automatically play. This feature can be activated through the manage songs section of the music page. Then, as you find potential fans, ask them to add your song to their own profile page. Baker even suggested making it into a contest, with everyone who participates being entered into the contest.

My Bulletin Space is a feature that allows members to post bulletins that appear on all their friends' pages for 10 days. Bulletins can only be posted on another person's page by people who are part of that person's network of friends, and they do not show up on the public profile page for that person. Artists will often use the feature to broadcast information about upcoming live shows. For MySpace members who are a part of an artist's group of friends, they will see these bulletins about upcoming shows when they log on to their MySpace page (MacPhearson, 2007).

Making regular entries to the blog is another way for artists to engage fans and keep them coming back to the MySpace page. Stories from the road are often interesting fodder for blogs, along with stories behind the songs and milestones in the artist's career. Make blogs personal by writing in the first-person tense. Ask fans to subscribe to the artist's blog, which they can do by clicking the link right after "latest blog entry" on the artist's page.

MySpace also features classified ads. This section gives musicians a chance to post classifieds, whether it's offering entertainment services, looking for a new drummer, or buying or selling musical equipment. Classifieds are sorted according to geographic location and subject area. Look

From	Date	Bulletin
Scott Thompson.	Apr 11, 2008 12:31 AM	wanna hear somethin funny?
The Last Straw	Apr 10, 2008 9:15 PM	Hey KNOXVILLE!! FRIDAY NIGHT @ BARLEY"S THE LAST STRAW &!!!!
The Dynamites feat. Charles Walker	Apr 10, 2008 3:31 PM	Dynamites Bring The Funk To 2nd Life In Half An Hour!
bruce burch	Apr 10, 2008 11:09 AM	Check out this event: The Dallas Austin Experience
The Dynamites feat. Charles Walker	Apr 10, 2008 10:16 AM	Dynamites headline Second Life Music Festival TODAY!
The Last Straw	Apr 9, 2008 6:20 PM	The Last Straw and Dirty Guvnah's this Friday N KNOXVILLE!!!
Mike	Apr 9, 2008 3:35 PM	I NEED HELP
The Last Straw	Apr 9, 2008 2:39 PM	HEY KNOXVILLE!!! We're coming to BARLEY"S!!!!
Kim	Apr 7, 2008 8:14 PM	NYC
Scott Thompson.	Apr 6, 2008 3:13 AM	man

post bulletin | view all

FIGURE 12.7
Example of My Bulletin Space with artist postings. (Printed with permission from MySpace.)

for the "classifieds" button on the top menu bar, usually the button on the far right. If geography is important, the next step is to click the "change city" button to the relevant city. Most major U.S. cities are listed. If the city of importance is not on the list, select the nearest major metropolitan area from the list. One of the subject options is Musician's XChange, which features musical equipment for sale and job opportunities.

Keywords: [   ]   search

Fri 04/11

High Energy Drummer Here Needing Home!!!! (Columbia, Nashville) by Josh B.

Thu 04/10

Punk drummer wanted (Nashville) by Johnny Carcinogen

Vocalist seeking brutal metal/hardcore band (Murfreesboro, Nashville) by Jack "Slut Killa" STAY PUFT

SEEKING ACOUSTIC GUITARIST FOR UP-COMING GIG!!!! (Nashville, TN) by Abby Fender

Lead Guitarist seeking band!! (Crossville, Knoxville, Chattanooga, Nashville) by J.J.

Drummer looking for gigs/tours (Nashville) by Weston(music page)

Bass player needed asap....... (Nashville) by Kevin New

Keyboardist/ backup vocals needed (knoxville,nashville) by jonathan bouren(Band Needed)

Touring Drummer Available (anywhere) by Rav

Looking for Acoustic Guitar Accompaniment (Nashville Area) by Ender Bowen

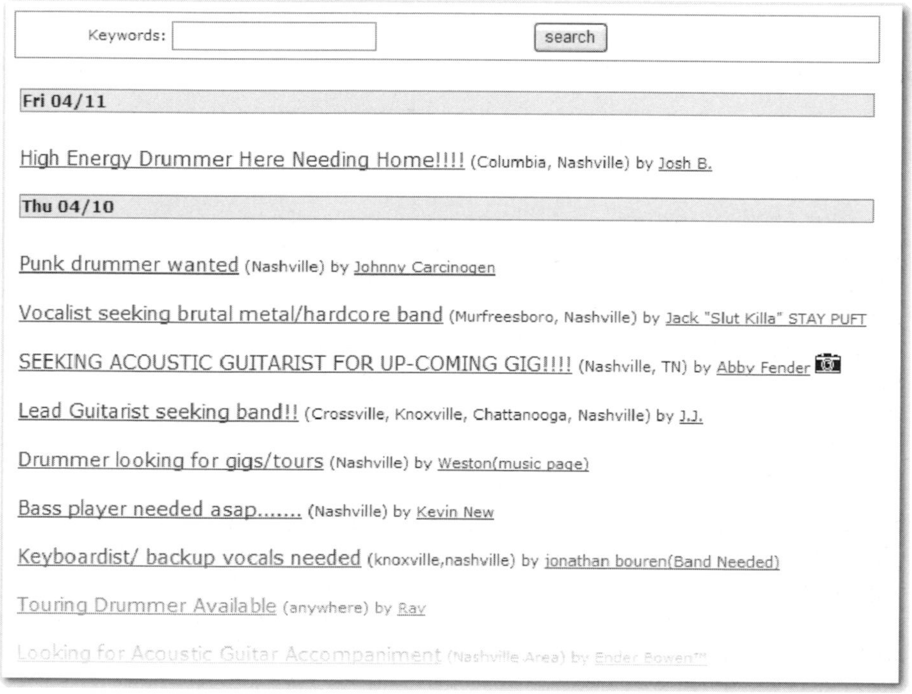

FIGURE 12.8
Classified ad postings on MySpace. (Printed with permission from MySpace.)

The classified section also allows users to search by keyword. Similar to Craigslist, the classifieds are listed by the date of posting, with the newest postings first. Artists and record labels are known to use MySpace to recruit members of the target market to work as street teams, either online or in the traditional sense.

EXAMPLE OF A SOCIAL NETWORKING STREET TEAM SOLICITATION

Groups » LEON HIGH SCHOOL!! » Topics » Teen Street Team Members Needed
Posted: Aug. 11, 2007 9:48 PM

Do you love attending exciting events? Do you like to be the center of attention whenever you enter a room? Are you interested in working in the music or fashion industry? If you answered yes, then you have what it takes to become a Blu Diamond! As a Blu Diamond you will be the life of the party at every Blu Diamond Event and best of all you're getting paid to be!

The Blu Diamonds are young ladies, ages 14–18, whose primary responsibilities include:

Promoting Blu Diamond Events—Hip Hop and R&B parties for teens ages 14–18 held every other month at local hotspots.

Assisting with the planning of upcoming events—Show your creativity by planning the hottest teen parties in town.

Attending monthly meetings to stay up-to-date—Our monthly meetings are fun as well as informative. Refreshments and raffle prizes are always provided by Blu Sapphire Promos.

Attendance at fundraisers such as car washes and sponsorship nights at local restaurants.—Our car washes and dinner fundraisers help us to raise money for things such as new t-shirts and other clothing items, the hottest DJs in Tallahassee, and to throw the hottest theme parties in town.

Attending Blu Diamond Events at local hotspots acting as either waitresses, photo crew, admissions, or hype girls.

We are currently in search of new girls between the ages of 14–18 to join the team. Please send a message AND add us as a friend if you are interested in becoming a Blu Diamond.

Street Team Members are paid based on the number of guests they bring to each Blu Diamond event. All promotional materials such as t-shirts, a web site, and flyers are provided to ensure your success. If you are interested in becoming a part of Tallahassee's underground music industry please send us a message with the following information:

Name
Cell Phone Number
Age
School
Grade

Favorite Hip Hop Artist
Favorite R&B Artist
Favorite Hangout Spot
Other Clubs/Sports you're a part of
Make sure to add us as a friend so that we can view your MySpace page. If your page
is private and we can view it we will not be able to respond to your message.

Permission granted from www.myspace.com/blu_sapphire.

MySpace also has a section called groups. These are "user groups" based on topic of interest and allow for members to find and interact with other members who share their interest in the group topic. User groups first appeared on the Internet early on in its development, with the first user groups dedicated to science and technology. Some of these groups were oriented around areas of research study at universities, and the groups allowed for like-minded people to exchange knowledge and ideas in the subject area of study. As the Internet grew, so did the variety of offerings for user groups. Soon, user groups were springing up on services like CompuServe and dedicated to almost any topic or area of interest, including computers, music, and science.

On MySpace, there are more than 300,000 music groups and more than 60,000 "Nightlife and Clubs" groups. Some of the groups are fan clubs for specific well-known artists, whereas others may be genre oriented. Peruse through the list of groups and read the descriptions of groups that may be relevant to promoting your artist. For example, the "Bay Area Music" group features musicians and bands from the San Francisco area.

FIGURE 12.9
MySpace example of a user group. (Printed with permission from MySpace.)

MySpace announced in April 2008 that MySpace Music would be expanded to enable artists to sell music downloads, concert tickets, and merchandise through their profile pages and ringtones through MySpace's sister company Jamba mobile service. The major labels are slated to get onboard and resolve previous copyright disputes with the social networking service (Veiga, 2008).

FACEBOOK

The Facebook social networking site has about 10% of the networking traffic as of late 2007. Originally available only to students with a school affiliation, Facebook opened its ranks to the public in September 2006 and doubled its traffic in one year (Hitwise). In May 2007, the site opened its platform, allowing software developers the chance to create integrated programs. From this, we are beginning to see more music-related applications. One such application dedicated to music is iLike. Born from garageband.com, iLike catalogs what users listen to through iTunes and other computer music players, compiles lists of recommendations, and allows users to share their music with friends.

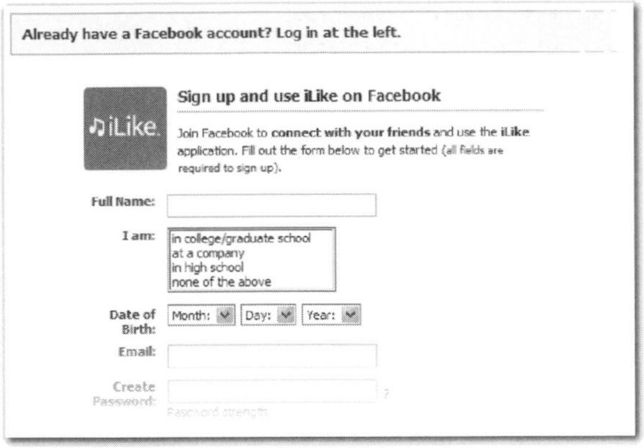

FIGURE 12.10
The iLike artist "add songs and events" page. (Courtesy of iLike.)

According to iLike, it promotes music democracy by exposing artists that users have not heard of but may like, according to their listening habits. Users download an iTunes sidebar that tracks their listening habits, which can be easily posted on their Facebook profile page.

For artists attempting to promote their music on Facebook through iLike, the iLike FAQ page offers the following tips:

1. Sign up on iLike and customize your artist page by adding your songs, concert schedule, and links. This will create an artist or band page for you on iLike. (The service is now set up to add iLike to your MySpace page.)
2. Ask your fans to sign up for iLike (at iLike.com or on Facebook.com) and "iLike" you. Then, ask them to add your music to their profile page. Then when their friends visit their page, it will expose them to your music. You can create a link guiding your visitors to the sign-up page. Go to your artist page on iLike.com and copy the links from the "Link to this page" section on the right-hand side of the page.
3. Have your fans install the iLike Sidebar, enabling them to document their listening patterns. If they listen to your music, and if enough of them do this using the iLike sidebar, your music will automatically work its way into the iLike music recommendation system and charts.

On Facebook, a person's iLike section shows up on their profile page with information on songs and artists they like and other music-related preferences. Artists on a person's profile list have the opportunity to post new events (mostly upcoming concert appearances) on the artist activity page.

YOUTUBE

YouTube is a video-sharing web site that opened in 2005; it has become very popular and created quite an impact in popular culture. Candid and bootleg videos have rocked political campaigns, spawned a new generation of videographers, launched new celebrities, and exposed others (i.e., the Michael Richards incident[1]). Artists have been exploiting YouTube to promote themselves and their videos, both professional and amateur. Much like video cameras produced a new genre of *America's Funniest Videos*, YouTube has become an outlet for everyone with camera. Internet marketer Dan Ackerman Greenberg has described the strategy behind using promotional techniques to increase the popularity of videos on YouTube.

[1] Michael Richards, who played Kramer on Seinfeld, was caught on video in a racially-tinged rant at a comedy club. One of the club's patrons used a cell phone to capture the moment and it was publicly disseminated, much to the dismay of Richards.

Both stars and emerging indie bands have found a home on YouTube for music videos. With the popularity of homemade videos, bands can create a music video or concert video at a fraction of what it would have cost just a few years ago. Postings on YouTube should be supported by a campaign to spread the word. Video "channels" can be created, and more traffic can be generated by joining forces with similar artists to set up a "channel" and promote it to the fans of all participating artists.

THE SECRET STRATEGIES BEHIND MANY "VIRAL" VIDEOS: A CASE STUDY

Dan Ackerman Greenberg

Have you ever watched a video with 100,000 views on YouTube and thought to yourself: "How did that video get so many views?" Chances are pretty good that this didn't happen naturally, but rather that some company worked hard to make it happen—some company like mine.

When most people talk about "viral videos," they're usually referring to videos like *Miss Teen South Carolina*, *Smirnoff's Tea Partay* music video, the *Sony Bravia* ads, *Soulja Boy*—videos that have traveled all around the Internet and been posted on YouTube, MySpace, Google Video, Facebook, Digg, blogs, etc. These are videos with millions and millions of views.

Here are some of the techniques to get at least 100,000 people to watch "viral" videos:

Secret # 1: Not All Viral Videos Are What They Seem

There are tens of thousands of videos uploaded to YouTube each day. (I've heard estimates between 10–65,000 videos per day.) I don't care how "viral" you think your video is; no one is going to find it and no one is going to watch it.

Our clients give us videos and we make them go viral. Our rule of thumb is that if we don't get a video 100,000 views, we don't charge. So far, we've worked on 80–90 videos and we've seen overwhelming success. In the past 3 months, we've achieved over 20 million views for our clients, with videos ranging from 100,000 views to upwards of 1.5 million views each. In other words, not all videos go viral organically. There is a method to the madness.

We've worked with: two top Hollywood movie studios, a major record label, a variety of very well known consumer brands, and a number of different startups, both domestic and international. This summer, we were approached by a Hollywood movie studio and asked to help market a series of viral clips they had created in advance of a blockbuster. The videos were 10–20 seconds each, were shot from what appeared to be a camera phone, and captured a series of unexpected and shocking events that required professional post-production and CGI [computer generated content]. Needless to say, the studio had invested a significant amount of money in creating the videos. But every time they put them online, they couldn't get more than a few thousand views.

(Continued)

We took six videos and achieved:

- 6 million views on YouTube
- ~30,000 ratings
- ~10,000 favorites
- ~10,000 comments
- 200+ blog posts linking back to the videos
- All six videos made it into the top 5 Most Viewed of the Day, and the two that went truly viral (1.5 million views each) were #1 and #2 Most Viewed of the Week.

The following principles were the secrets to our success.

2. Content Is NOT King

If you want a truly viral video that will get millions of people to watch and share it, then yes, content is key. But good content is not necessary to get 100,000 views if you follow these strategies.

Don't get me wrong: Content is what will drive visitors back to a site. So a video must have a decent concept, but one shouldn't agonize over determining the best "viral" video possible. Generally, a concept should not be forced because it fits a brand. Rather, a brand should fit into a great concept. Here are some guidelines to follow:

- *Make it short.* 15–30 seconds is ideal; break down long stories into bite-sized clips
- *Design it for remixing.* Create a video that is simple enough to be remixed over and over again by others. Ex: *Dramatic Hamster*
- *Don't make an outright ad.* If a video feels like an ad, viewers won't share it unless it's really amazing. Ex: *Sony Bravia*
- *Make it shocking.* Give a viewer no choice but to investigate further. Ex: *UFO Haiti*
- *Use fake headlines.* Make the viewer say, "Holy crap, did that actually happen?!" Ex: *Stolen NASCAR*
- *Appeal to sex.* If all else fails, hire the most attractive women available to be in the video. Ex: *Yoga 4 Dudes*

These recent videos would have been perfect had they been viral "ads" pointing people back to web sites:

- Model Falls in Hole on Runway
- Cheerleader Gets Run Over by Football Team
- PacMan: The Chase
- Dude
- Dog Drives Car
- Snowball—Dancing Cockatoo

3. Core Strategy: Getting onto the "Most Viewed" Page

Now that your video is ready to go, how is it going to attract 100,000 viewers?

The core concept of video marketing on YouTube is to harness the power of the site's traffic. Here's the idea: Something like 80 million videos are watched each day on YouTube, and a significant number of those views come from people clicking the "Videos" tab at the top. The goal is to get a video on that Videos page, which lists the Daily Most Viewed videos.

If you succeed, the video will no longer be a single needle in the haystack of 10,000 new videos per day. It will be one of the 20 videos on the Most Viewed page, which means that you can grab 1/20th of the clicks on that page! And the higher up on the page your video is, the more views you are going to get.

So how do you get the first 50,000 views you need to get your videos onto the Most Viewed list?

- *Blogs.* Reach out to individuals who run relevant blogs and actually pay them to post your embedded videos. Sounds a little bit like cheating/PayPerPost, but it's effective and it's not against any rules.
- *Forums.* Start new threads and embed your videos. Sometimes, this means kick starting the conversations by setting up multiple accounts on each forum and posting back and forth between a few different users. Yes, it's tedious and time consuming, but if you get enough people working on it, it can have a tremendous effect.
- *MySpace.* Plenty of users allow you to embed YouTube videos right in the comments section of their MySpace pages. Take advantage of this.
- *Facebook.* Share, share, share. Take Dave McClure's advice and build a sizeable presence on Facebook, so that sharing a video with your entire friends list can have a real impact. Other ideas include creating an event that announces the video launch and inviting friends, writing a note and tagging friends, or posting the video on Facebook Video with a link back to the original YouTube video.
- *Email lists.* Send the video to an email list. Depending on the size of the list (and the recipients' willingness to receive links to YouTube videos), this can be a very effective strategy.
- *Friends.* Make sure everyone you know watches the video and try to get them to email it out to their friends, or at least share it on Facebook.

Each video has a shelf life of 48 hours before it's moved from the Daily Most Viewed list to the Weekly Most Viewed list, so it's important that this happens quickly. When done right, this is a tremendously successful strategy.

4. Title Optimization

Once a video is on the Most Viewed page, what can be done to maximize views? It seems obvious, but people see hundreds of videos on YouTube, and the title and thumbnail are an easy way for video publishers to actively persuade someone to click on a video. Titles can be changed a limitless number of times, so have a catchy (and somewhat misleading) title for the first few days, then later switch to something more relevant to the brand. Recently, I've noticed a trend toward titling videos with the phrases "exclusive," "behind the scenes," and "leaked video."

(Continued)

5. Thumbnail Optimization

If a video is sitting on the Most Viewed page with 19 other videos, a compelling video thumbnail is the single best strategy to maximize the number of clicks the video gets.

FIGURE 12.11
YouTube example of a video thumbnail (permission granted by YouTube and artist Grant Peeples).

YouTube provides three choices for a video's thumbnail, one of which is grabbed from the exact middle of the video. As you edit your videos, make sure that the frame at the very middle is interesting. It's no surprise that videos with thumbnails of half-naked women get hundreds of thousands of views. Not to say that this is the best strategy, but you get the idea. Two rules of thumb: The thumbnail should be clear (suggesting high video quality) and ideally it should have a face or at least a person in it.

Also, when you feel particularly creative, optimize all three thumbnails, then change the thumbnail every few hours. This is definitely an underused strategy, but it's an interesting way to keep a video fresh once it's on the Most Viewed list.

6. Commenting: Having a Conversation with Yourself

Every power user on YouTube has a number of different accounts. So should you. A great way to maximize the number of people who watch your videos is to create some sort of controversy in the comments section below the video. Get a few people in your office to log in throughout the day and post heated comments back and forth (you can definitely have a lot of fun with this). Everyone loves a good, heated discussion in the comments section—especially if the comments are related to a brand/startup.

Also, don't be afraid to delete comments. If someone is saying your video (or your startup) sucks, just delete their comment. Don't let one user's negativity taint everyone else's opinions.

We usually get one comment for every 1,000 views, since most people watching YouTube videos aren't logged in. But a heated comment thread (done well) will engage viewers and will drive traffic back to your sites.

7. Releasing All Videos Simultaneously

Once people are watching a video, how do you keep them engaged and bring them back to a web site?

A lot of the time our clients say: "We've got five videos and we're going to release one every few days so that viewers look forward to each video."

This is the wrong way to think about YouTube marketing. If you have multiple videos, post all of them at once. If someone sees your first video and is so intrigued that they want to watch more, why would you make them wait until you post the next one?

Give them everything up front. If a user wants to watch all five of your videos right now, there's a much better chance that you'll be able to persuade them to click through to your web site. Don't make them wait after seeing the first video, because they're never going to see the next four.

Once your first video is done, delete your second video, then re-upload it. Now you have another 48-hour window to push it to the Most Viewed page. Rinse and repeat. Using this strategy, you give your most interested viewers the chance to fully engage with a campaign without compromising the opportunity to individually release and market each consecutive video.

8. Strategic Tagging: Leading Viewers Down the Rabbit Hole

YouTube allows you to tag your videos with keywords that make your videos show up in relevant searches. For the first week that your video is online, don't use keyword tags to optimize the video for searches on YouTube. Instead, you can use tags to control the videos that show up in the Related Videos box. Why?

I like to think about it as leading viewers down the rabbit hole. The idea here is to make it as easy as possible for viewers to engage with all your content, rather than jumping away to "related" content that actually has nothing to do with your brand/startup.

So how do you strategically tag? Choose three or four unique tags and use only these tags for all of the videos you post. I'm not talking about obscure tags; I'm talking about unique tags—tags that are not used by any other YouTube videos. Done correctly, this will allow you to have full control over the videos that show up as "Related Videos." When views start trailing off after a few days to a week, it's time to add some more generic tags, tags that draw out the long tail of a video as it starts to appear in search results on YouTube and Google.

9. Metrics/Tracking: How to Measure Effectiveness

The following is how to measure the success of your viral videos.

For one, tweak the links put up on YouTube (whether in a YouTube channel or in a video description) by adding "?video=1" to the end of each URL. This makes it much easier to track inbound links using Google Analytics or another metrics tool.

TubeMogul and VidMetrix also track views/comments/ratings on each individual video and draw out nice graphs that can be shared with the team. Additionally, these tools follow the viral spread of a video outside of YouTube and throughout other social media sites and blogs.

Conclusion

The Wild West days of *Lonely Girl* and *Ask A Ninja* are over. You simply can't expect to post great videos on YouTube and have them go viral on their own, even if you think you have the best videos ever. These days, achieving true virality takes serious creativity, some luck, and a lot of hard work. So, my advice: Fire your PR firm and do it yourself.

This was written by Dan Ackerman Greenberg, co-founder of viral video marketing company. The Comotion Group and lead TA for the Stanford Facebook Class. Dan graduated from the Stanford Management Science & Engineering Masters program in June 2008. www.thecomotiongroup.com. Reprinted by permission of author. www.techcrunch.com/2007/11/22/the-secret-strategies-behind-many-viral-videos.

IMEEM

Imeem touts itself as "an online social network where millions of fans and artists discover new music, videos, and photos, and share their tastes with friends." As of February 2008, it was drawing 24 million visitors per month by specializing in free streaming music and music-video playlists. Imeem also has 100 million unique users in their widget network. Using imeem widgets, you can embed your playlist on any third-party site. They offer single click buttons, and the embedding code so musicians can add it to their own web site, blog, and so on.

Members have the opportunity to set up a profile, create playlists, and share these lists with other members who can listen to these playlists. Users can create, recommend, and discover music, film, art, and pop culture while connecting with other members with common interests, referred to as "meems" (Harbin, 2007).

With imeem, you can add a song to your profile on MySpace, Facebook, BlackPlanet, Tagged, and Hi5 (and the site keeps adding others). Through RSS syndication, content can be updated on these external sites. Artists are able to upload content via the upload link in the upper right corner of the masthead. Imeem is heavily focused on media, especially music, and its users can build playlists and explore new music easily. Imeem has also worked out agreements with the four major labels to license full-version tracks of their music for streaming. In exchange, the labels receive a portion of the advertising revenue.

To create an artist account, first you must register to create an account by clicking on the "Create Your Account" button next to the "Log In" button. Account registration requires your name, e-mail address, a password, and date of birth. To upgrade your profile to an artist's profile, you must check your e-mail for the verification code. When that e-mail arrives, click on the link provided to verify the code, or if that doesn't work, copy the code and log on to www.imeem.com/v.aspx. Once registered, you can begin to develop the artist's profile via the "account settings" feature. One of the tabs reads "upgrade profile," and this is where you have the opportunity to upgrade the account to musician status.

FIGURE 12.12
Imeem song page with many communication features. (Permission granted from imeem.)

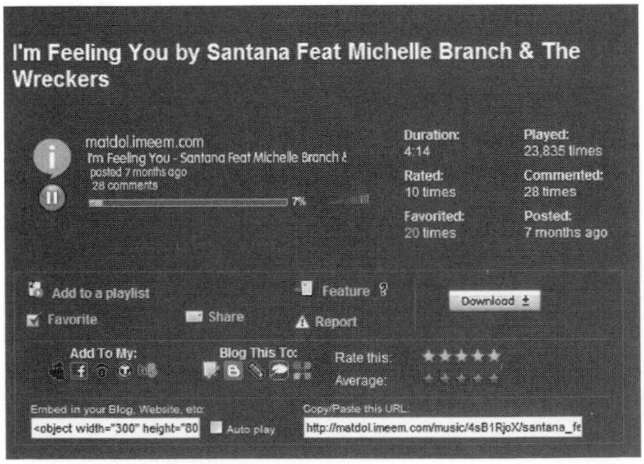

With imeem, you have the opportunity to develop the artist's profile, including photos, font styles, theme, and layout. All aspects of the page can be edited or adjusted, including the page header, section headers, wallpaper background, and general page settings.

The profile layout has areas for posting upcoming events, bio (about) music, playlists, friends and groups, and recent media.

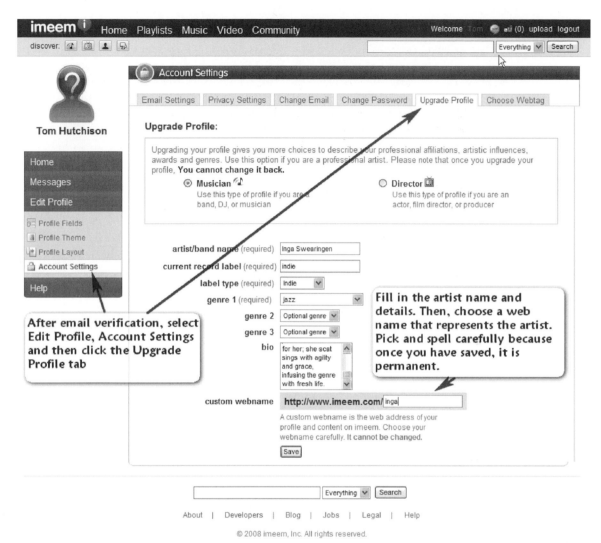

FIGURE 12.13
Imeem upgrade to musician account (permission granted from imeem).

The song load section allows for uploading an album cover, song title, artist, album, description, and tags. Imeem supports either MP3, OGG, WMA, or M4A formats. The only place to add HTML code to imeem profiles is in the "profile comments" box. Once the HTML is posted, the correlating images and links appear. If the HTML code is not interpreted correctly for some items, such as pulling images from other sites, it may be necessary to adjust and preview until everything is in order.

Imeem also allows for analyzing user traffic with the "view stats" feature. You can see how many hits each item gets.

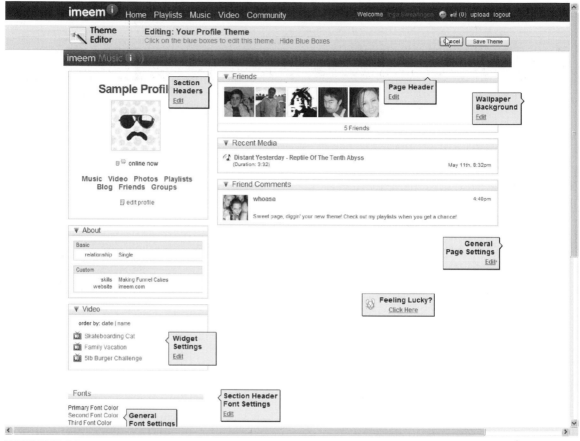

FIGURE 12.14
Page layout for imeem (permission granted from imeem).

BEBO

Another social networking site, Bebo, is not as heavy on music but does offer some music features. Bebo describes itself as a social media network. Bebo is similar to imeem in that an artist's personal profile and band profile are separate (Harbin, 2007). Band profiles provide different features that help promote the artist, such as tour dates and list of fans. Users can add their AIM, Skype, and Windows Live Messenger user names to their profiles. Members also have the opportunity to create fan groups, to support their favorite artists, and create playlists for other members to listen to.

First, you must sign up for a user account. Under profile information, you can add the band's bio or other personal information. After the account is created, that becomes the personal home page. After the profile is completed on the artist's "home" page, click on "Music" to build a musician or band profile. Under

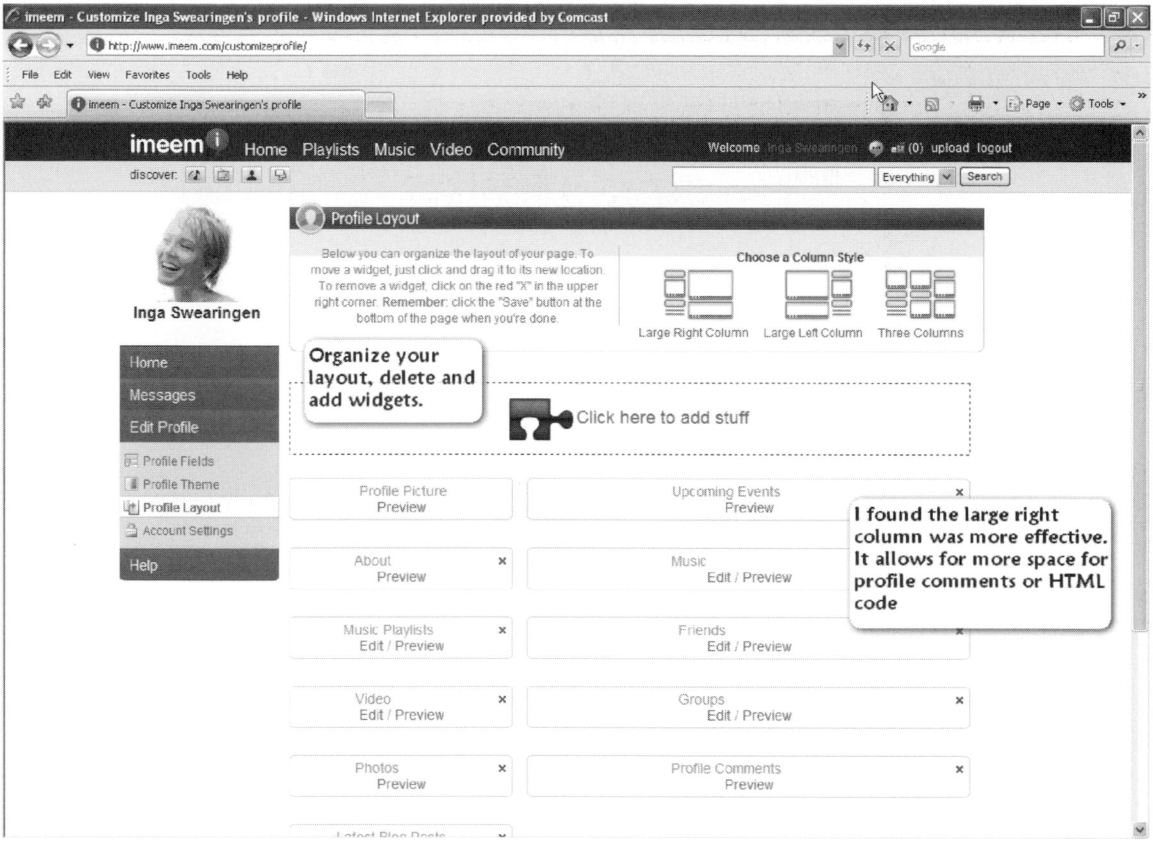

FIGURE 12.15
Imeem layout for an artist account (permission granted from imeem).

the music tab, you can select "register your band," which loads a submission page called register your music, which allows you to upload your music.

When the "register band" fields have been completed, click on the image under the caption "Band member of" to start creating the band profile, including selecting photos, skins, and a background layout. The photo tab allows you to upload photos, music, and videos.

Bebo has profile modules that you can use and arrange to provide features for the page. The modules include band members, fans, video box, polls, blog, upcoming shows, music store, songs, music albums, photos, comments, white board, wall, widgets, and my skins. For the "upcoming shows" feature, you are prompted to enter dates, times, duration, location, and details on each show. Bebo provides "web badges," which are graphical widgets that can be pasted on other sites and blogs directing web surfers to the artist's Bebo page.

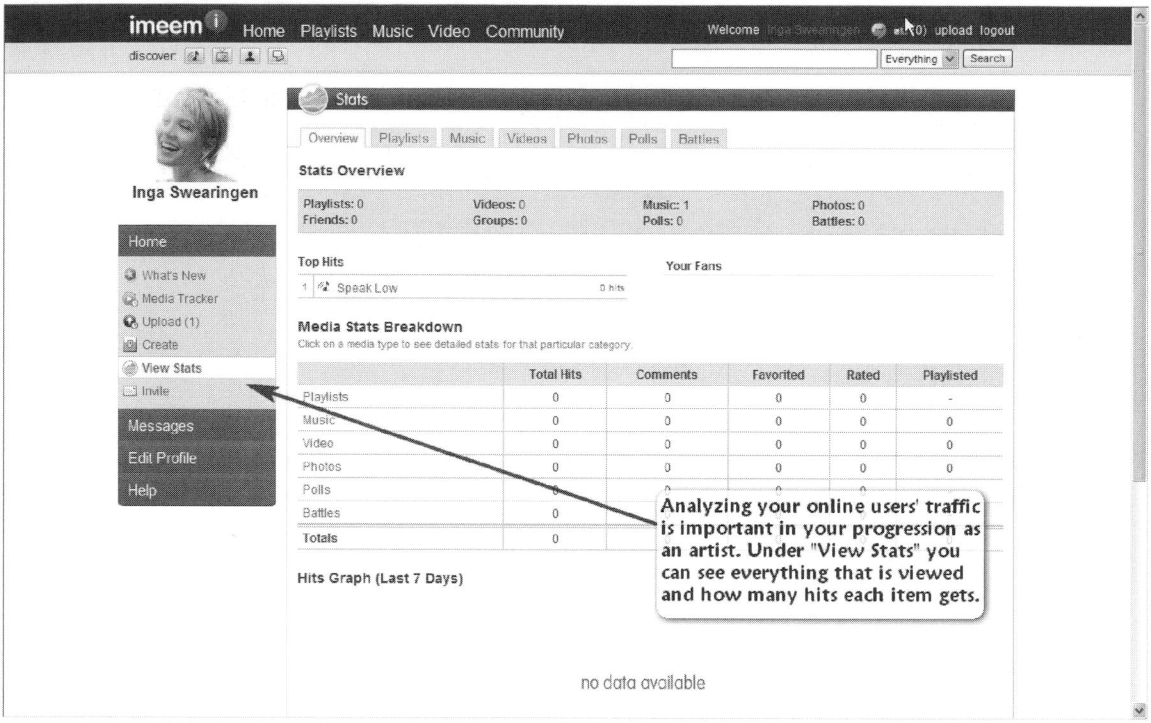

FIGURE 12.16
Imeem statistics (permission granted from imeem).

OTHER SOCIAL NETWORKING SERVICES

There are plenty of other social networking services and several upstarts that focus on music. Some of these sites may take off in popularity, whereas others may falter.

- *BlackPlanet.* This is a social networking site focused on African Americans, offering networking and news about what's hot in music, fashion, sports, and events in the black community. The network offers music features including streaming, creating playlists and having members add their favorite artist to their profile. BlackPlanet is also connected through imeem, so that songs previewed there can be added to the user's BlackPlanet profile.
- *Hi5.* This is another social networking service with a large music component. Their web site calls it "the place where young people come to stay in touch with friends, meet new people, create & explore content, and express themselves. hi5 provides a platform for established artists, underground talent, and everyday people to all gain prominence amongst a worldwide

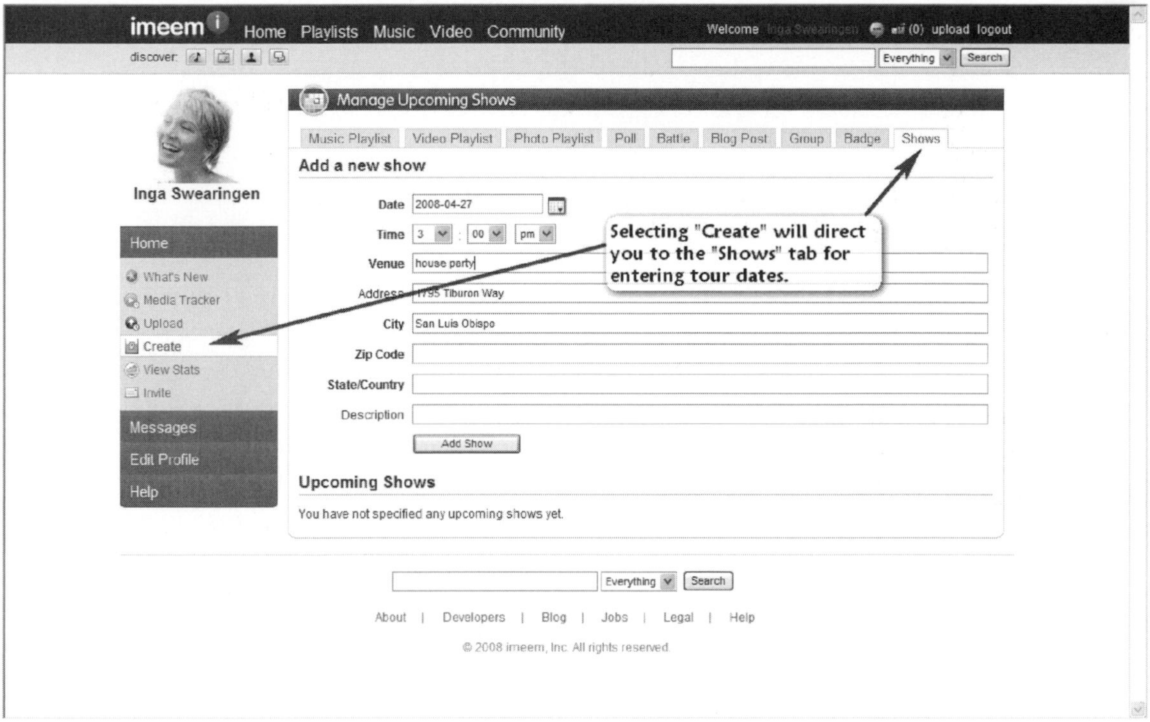

FIGURE 12.17
Imeem upcoming shows feature (permission granted from imeem).

audience. Our members voice their opinions and increase the significance of an artist, a person, or a piece of content."

- *MOG.* MOG connects users by recommendations and by matching those with similar tastes in music. It has direct links to Amazon and iTunes for music purchases. It has a widget that tracks the user's listening habits on iTunes. MOG has features to create artist profiles.
- *Music.com.* This online community (still in the beta stage at the time of this writing) provides for user rating and creation of favorites lists. It also features artist profiles.
- *ReverbNation.* Provides a platform for artists to feature their music, profiles, tour schedules and so forth, and share this information with fans and industry people. The service features widgets to add the artist's profile to popular social networks.

With so many new social networking services cropping up, no one is certain what will be the "next big thing." An astute marketing manager learns to stay abreast of the changing trends in social networking and respond quickly and comprehensively to changes in the marketplace for social interaction.

GLOSSARY

Blogging or blog – Short for weblog, it is a personal online journal that is frequently updated and intended for general public consumption.

Hypertext markup language (HTML) – The predominant authoring language for the creation of web pages. HTML defines the structure and layout of a web document by using a variety of tags and attributes.

Social networking – The practice of interacting with and expanding the number of one's business or social contacts by making connections typically through social networking web sites such as MySpace and Facebook.

URL – Uniform Resource Locator, the global address of documents and other resources on the Web. The first part of the address is called a protocol identifier, and it indicates what protocol to use; the second part is called a resource name, and it specifies the IP address or the domain name where the resource is located.

Viral marketing – A marketing phenomenon that uses online social networking to facilitate and encourage people to pass along a marketing message voluntarily and exponentially. Sometimes the marketing message is imbedded in, or attached to, a message that participants find interesting and are willing to forward to others.

Web widgets – A widget is a small application that can be ported to and run on different web pages by a simple modification of the web page's HTML.

WYSIWYG – Pronounced WIZ-zee-wig, short for "what you see is what you get." A WYSIWYG application enables you to see on the display screen exactly what will appear when the document is printed or published to the Web.

REFERENCES AND FURTHER READING

Baker, Bob. (2006, July 1). MySpace music marketing: How to promote & sell your music on the world's biggest networking web site. http://www.bob-baker.com/buzz/bobbooks.html.

Boyd, Danah, and Ellison, Nicole. (2007). Social Networking sites: Definition and conception, www.danah.org/papers/worksinprogress/SNSHistory.html.

Brothers, Patrick, and Layton, Julia. (2007). How MySpace works, http://computer.howstuffworks.com/myspace.htm.

Cio-today. (2008, January 30). Could social networking save the music industry, www.cio-today.com/story.xhtml?story_id=011001CRJKFU.

Dawes, John, and Sweeney, Tim. (2000). The Complete Guide to Internet Promotion for Artists, Musicians and Songwriters. Tim Sweeney and Associates, Temecula, CA.

Greenberg, Dan Ackerman. The secret strategies behind many "viral" videos: A case study. www.techcrunch.com/2007/11/22/the-secret-strategies-behind-many-viral-videos.

Harbin, Jennifer. (2007). Universal Music Group new media intern guide. Unpublished manuscript.

Hitwise. (2007, March 14). IMeem and Bebo are the fastest movers, www.hitwise.com/press-center/hitwiseHS2004/socialnetworkingmarch07.php.

MacPhearson, Michelle. (2007, February 5). Viewing and posting bew Myspace. Bulletins, www.articles3000.com/Internet/60055/Viewing-And-Posting-New-Myspace-Bulletins.html.

Mills, Elinor, Advertisers look to grassroots marketing, http://news.com.com/2100-1024_3-6057300.html.

Nevue, David. (2005). How to Promote Your Music Successfully on the Internet. Midnight Rain Productions. Eugene, OR.

Veiga, Alex. (2008, April 4). MySpace will offer musicians a new outlet. *The Tennessean*, section E, p. 1.

Vincent, Frances. (2007). *MySpace for Musicians*. Thomson Course Technology, Boston, MA.

CHAPTER 13
Internet Obstacles

The worst mistake an Internet newbie can make is quitting his day job. The Web is filled with how-to books that promise a fortune of riches and success to those who buy and follow the book. Although the Internet is full of opportunities to engage in commerce, competition is fierce. This chapter addresses some of the Internet marketing no-nos, pitfalls, and outdated or useless strategies that may still be touted online. The Web is full of top-10 lists of things to avoid, mistakes, and no-nos. Maybe the first item to mention should be "Leave it to Letterman and give us a break on the top-10 lists." But here goes.

WEB SITE MISTAKES
Launch When Ready

One of the worst mistakes a web designer can make is to launch a web site before it is ready and has undergone beta testing. Check for browser compatibility, spelling errors, broken links, missing components, and so on before telling the world about your site. That includes signing up with search engines. You should do that only after the site is in good working order. Ask friends and family to "test drive" the web site and report back any irregularities.[1]

Oops! I Did It Again

Another mistake, described by author Robyn Tippins in *What Not to Do on Your Website*, is "do not post things you will regret." Sometimes, words posted on

---

[1] It's a good idea to also invest in a server plan that allows you to set up sub-domains (or have your developer do it on their site) and have the sub-domain be the place that testing is done. It's never good to test on a live domain, even if you don't announce the site to anyone. This is also good moving forward to allow for testing of upgrades to the site.

a web site or blog can come back to haunt the author, just as public comments made by politicians can cause a backlash. And with the printed word on a web site, it's even more difficult to deny it because there it is in writing. College graduates are finding that their shenanigans posted on MySpace are hindering in the job search. All comments, text, photos, e-mail replies, and the like should be carefully critiqued as to their impact before hitting that submit button.

Go Lightly on the Ads

Too many ads clutter up a web site, compete for the viewer's attention, and drive people away. A professional site should concentrate on the product for which the site is created, instead of becoming a billboard for selling ad space. Tippens suggested placing ads beneath a story or blog, in the left sidebar, or "wherever the eye naturally falls."

With the introduction of Google AdSense, many web sites that were formerly ad-free have become littered with ads in an effort for the site owner to profit from maintaining the site. Many of these sites used to be a "labor of love," but they are now commercial enterprises. That is fine for the hobbyist, but professional music business entities, including artists, labels, and affiliated music business web sites, should not have such advertising messages on the site except those that promote the products related to the purpose of the site.

In his book *How to Promote Your Music Successfully on the Internet,* author David Nevue stated that banner ads can actually hurt your promotional efforts. There are many companies now on the Internet that foster banner exchanges, much like the link exchanges mentioned in Chapter 11. Among the reasons to stay away from such arrangements are (1) they lure visitors away from your site before they've had time to explore it, (2) they are unsightly and detract from the site, and (3) they can often be poorly targeted for products and services that have nothing to do with your site.

Put Function before Form

Don't let all the bells and whistles take priority over the purpose of the site. The visual aspect of the site should support the site's function, not detract from it. In the music industry, it is common to find music samples on web sites, but visitors should be given the option to turn off the music if it is detracting from their mission or slowing down the process.[2]

---

[2] Or, better yet, make them turn on the music if they want to hear it. If they really want to hear the music, they'll be happy to click a button to play it. One of the problems with streaming music when a page loads is that, if the user is on dial-up or a low-speed DSL connection or any type of slow connection, loading and playing the stream is going to take away from the loading and operation of the visual aspects of the page and the user won't know why until the music starts playing, which may not be right away. They'll just bail out of the page before they even get to see all of it and never return.

Table 13.1	36 Website Mistakes That Can Kill Your Business
Poor load time	Multiple banners and buttons
Poor overall appearance	Use of frames
Spelling/grammar mistakes	Large fonts
No contact information	Popup messages
Poor content	Over use of Java
Poor navigation	Poor use of tables
Broken links and graphics	Poor organization
Poor browser compatibility	Poor use of mouse-over effects
Large, slow-loading graphics	Overpowering music set to autoplay
Too many graphics	Too much advertising
Pages scrolling to oblivion	Large welcome banners
Multiple use of animated graphics	Multiple-colored text
Animated bullets	Text difficult to read
Too many graphic or line dividers	Multiple use of different fonts
Different backgrounds on each page	Confusing or broken links
Busy, distracting backgrounds	Under construction signs
Confusing	Bad contrast between text and background
No meta tags	Large scrolling text across the page

Source: Adapted from Shelley Lowery.

WEB MARKETING MISTAKES

In an attempt to recruit more business, commercial Internet marketing firms are generally more than willing to point out common marketing mistakes made by amateurs. Many have articles on the top 5 or top 10 mistakes commonly made in Internet marketing. The following is a composite of those mistakes most likely to impact web marketing for musicians and their business entities. Many of these mistakes are merely omissions of important marketing recommendations found elsewhere in this book.

Too Much Flash

One common theme among the lists is using slow-loading programs, too much flash, and making visitors wait while these programs load. This mistake was the first listed in several articles. An article on the AllBusiness site stated that if you

FIGURE 13.1
This site boasts that it is the world's worst site. (Reprinted with permission: hwww.angelfire.com/super/badwebs.)

make viewers wait too long, "you can say goodbye to your potential customers." You need to build fast-loading web pages.

Too Much Emphasis on Design

Another commonly listed mistake is focusing too much on design. In his article "Five Big Online Marketing Mistakes," author Ian Lurie stated the mistake is in "thinking cool design equals good marketing." He stated that there is a difference between using the right design to appeal to customers and the cool design. The design should reflect the tastes of the customer, not the company CEO.

Depending Too Much on Search Engines

Another commonly mentioned mistake had to do with search engine optimization (SEO) (see Chapter 7). Experts advise not to depend too heavily on SEO to draw traffic but also not to neglect submitting to search engines and optimizing. Lurie stated, "too many consultants are running around telling companies that they can achieve nirvana by simply changing meta-tags." An article on ShoestringBranding.com stated, "if we find ourselves spending too much time fiddling with page titles, changing the wording in our headlines or worrying about our keyword density, it is time to stop and ask ourselves if that time wouldn't be better utilized creating useful content and building loyal customers."

Failing to Collect Information

From e-mail addresses to traffic information, consultants recommend that you get as much feedback and contact information from your visitors as is reasonable. You need to know as much as possible about your visitors and your market to be effective at marketing. However, moderation is the key in requiring visitors to divulge information about themselves (see the following paragraph).

Complexity

Another commonly mentioned faux pas deals with how hard visitors must work to find what they are looking for. In his article "Do You Make These Internet Marketing Mistakes?" Peter Geisheker noted that it is annoying to try to navigate sites "where you have to click on generic graphics or parts of a flash picture to try and find the link to the information you are looking for." His advice is to keep it simple. The AllBusiness article warns against "giving users the third degree" by making them jump through hoops or answer a battery of questions to move ahead with the transaction. The article stated that every question "beyond name and email address will cost you 10% to 15% of your potential customers."

Not Using Viral Marketing

Advertising and search engine listing are important, but in the music business, the power of word of mouth (WOM) is overwhelming. Viral marketing achieves this. AllBusiness suggests the "tell-a-friend" option online and the use of branded T-shirts that your customers can wear for offline WOM marketing. The web site e-consultancy offers several reasons why viral marketing campaigns fail. Among them are the following:

1. *Neglecting seeding.* It is important to take advantage of mailing lists, press releases, forums, and other outlets where the marketing message can be placed.
2. *Failing to create an incentive for users to pass it along.* What good is a marketing message that is not passed along? It's not viral if it doesn't spread. Attach your message to something that is likely to be forwarded. Eighty-eight percent of web users say they have forwarded jokes.
3. *Trying to copy a popular viral campaign.* What works in one situation may not translate to your specific needs.
4. *Failing to integrate viral campaigns with other marketing efforts.* The integration of online and offline promotions is covered in another chapter in this book.

The AllBusiness article stated that "no one is online all the time. To successfully market an Internet site, you need to market offline, too."

5. *Using a sledgehammer instead of a scalpel.* Often, simple ideas—such as e-mail signatures—produce better results than more complex campaigns.

6. *Forgetting to ask the user to take action.* The success of a viral campaign depends on the recipients actually doing something, such as visiting your web site, voting for your artist in a contest, or purchasing the product.

INTERNET MARKETING ETHICS

Unethical business practices seem to be of major concern lately. Ever since Enron first toppled, consumers and law enforcement have been more diligent when dealing with companies. We hear of various scams being perpetrated on the Internet, and even rumors about scams. It makes it hard to tell what is legitimate and what is fly-by-night. Anyone dealing with e-commerce needs to understand consumers' reluctance to give out personal information, especially credit card, banking, or social security numbers. Predatory computer hackers roam the Internet causing problems and spreading malicious software. (Ironically enough, while writing this section I was attacked by the Trojan virus.)

Misrepresentation

Deliberate misrepresentation that causes another person to suffer damages, usually monetary losses, is fraud. There have been elements within the recording industry in the past who have crossed the line and been subjected to prosecution or lawsuits. *Payola* is a prime example, where the popularity of songs heard on the radio was not based on decisions made by program directors but instead was based on how much payment was exchanged in consideration of airplay. The payola law does not make it illegal to pay for airplay, but it states that such payment must be disclosed to listeners. Why is it then so important to record labels that they not disclose that payment was made to secure airplay? It is because they want the public to believe the song is only being played because that is what the public requests to hear.

Other areas of marketing have existed in the gray area of ethics to where a certain amount of deception is now considered standard. When "nine out of ten doctors recommend" a certain product, you have to wonder who those nine doctors are and who they work for. In advertising, celebrity spokespeople are hired and paid to endorse products, product features are enhanced, photographs are manipulated to show products in the best light, and press releases are created to spin information in favor of the company sponsoring the release.

Having said that, a certain amount of perception is fabricated and manufactured by marketers to give the appearance of a naturally occurring rise in popularity for certain cultural products. Specifically, record labels and artist managers hire people to pose as fans to create a word-of-mouth buzz about the artist. Internet marketers are routinely charged with the task of creating multiple fictitious profiles on social networking sites for the purpose of posing as fans and promoting the artist. It is not that different from hiring celebrity spokespeople or actors to pose as consumers to tout products in advertisements. The ethical dilemma in both situations comes from the fact that there is no disclosure of who these spokespeople really are. The Internet makes it even more difficult for consumers to determine who they are getting information from because profiles can be easily falsified. Web 2.0 has created an atmosphere of user-generated content that is not subjected to the usual fact checking, editing, and accountability. Wikipedia and MySpace have been dealing with these issues for the past few years; other social networking sites may start employing more stringent methods of verification as these problems emerge. But for now, the Web is still the Wild West when it comes to integrity and accountability. A certain amount of consumer skepticism is to be expected.

Scamming, Phishing, and Pharming

There is no shortage of scam artists on the Internet, employing all sorts of techniques to separate honest people from their money. The most notorious is the Nigerian 419 e-mail scam. It preys on people's sense of greed by purporting to have a large sum of inheritance money to give to the recipient, but only after the person pays some transfer fees. These fees start out small, but once these scammers hook a victim, they keep going back for more fees. The unsuspecting victim then agrees to pay the other fees in an attempt to recoup the previous fees they have paid by claiming their "prize." (See Figure 13.2.)

Other scams—called *phishing* and *pharming*—are not as notorious, are more personal in nature, and are designed to elicit private financial information that the scammers can use for identity fraud. Here is how phishing works: Internet fraudsters send spam e-mail messages that direct the victim to a website that looks just like a legitimate organization's site, typically a bank or PayPal. The site is not the official bank site but a bogus site that looks like it could be real.

It is designed to trick you into divulging your personal financial information by telling you that there is a problem with your bank account and you need to enter your password and other identification numbers to rectify the problem. The e-mail will contain links to the bank's sites, with the legitimate bank domain name in the link, but the HTML code directs the victim to a different site. HTML code for links can be written as <ahref="www.scam-operators-fake-site.com"> www.ReputableBank.com</a>. Note that the web address is different than the text for the link (see Figure 13.3).

Pharming is the latest threat. In the article "Privacy and the Internet: Traveling in Cyberspace Safely" (Privacy rights clearinghouse), pharming is described as "criminals' response to increasing awareness about phishing." Consumers have been alerted to the presence of phishing schemes and have been warned that, when receiving a message that purports to be from your bank, the safest option is to type in the web address of that site rather than click on the embedded links in the e-mail. Pharming works by redirecting the visitor after they have gone to the legitimate site. Presently, it is only possible for the scam artists to instigate on "http" sites, not on the more secure "https" (s for secure) pages.

DEAR SIR,

URGENT AND CONFIDENTIAL BUSINESS PROPOSAL

I AM MARIAM ABACHA, WIDOW OF THE LATE NIGERIAN HEAD OF STATE, GEN. SANI ABACHA.
AFTER HE DEATH OF MY HUSBAND WHO DIED MYSTERIOUSLY AS A RESULT OF CARDIAC ARREST, I
WAS INFORMED BY OUR LAWYER, BELLO GAMBARI THAT, MY HUSBAND WHO AT THAT TIME WAS THE
PRESIDENT OF NIGERIA, CALLED HIM AND CONDUCTED HIM ROUND HIS APARTMENT AND SHOWED
HIM FOUR METAL BOXES CONTAINING MONEY ALL IN FOREIGN EXCHANGE AND HE EQUALLY MADE
HIM BELIEVE THAT THOSE BOXES ARE FOR ONWARD TRANSFER TO HIS OVERSEAS COUNTERPART FOR
PERSONAL INVESTMENT.

ALONG THE LINE, MY HUSBAND DIED AND SINCE THEN THE NIGERIAN GOVERNMENT HAS BEEN
AFTER US, MOLESTING, POLICING AND FREEZING OUR BANK ACCOUNTS AND EVEN MY ELDEST SON
RIGHT NOW IS IN DETENTION. MY FAMILY ACCOUNT IN SWITZERLAND WORTH US$22,000,000.00
AND 120,000,000.00 DUTCH MARK HAS BEEN CONFISCATED BY THE GOVERNMENT. THE GOVERNMENT
IS INTERROGATING HIM (MY SON MOHAMMED) ABOUT OUR ASSET AND SOME VITAL DOCUMENTS. IT
WAS IN THE COURSE OF THESE, AFTER THE BURIAL RITE AND CUSTOMS, THAT OUR LAWYER SAW
YOUR NAME AND ADDRESS FROM THE PUBLICATION OF THE NIGERIAN BUSINESS PROMOTION
AGENCY. THIS IS WHY I AM USING THIS OPPORTUNITY TO SOLICIT FOR YOUR CO-OPERATION AND
ASSISTANCE TO HELP ME AS A VERY SINCERE RESPONSIBLE PERSON. I HAVE ALL THE TRUST IN
YOU AND I KNOW THAT YOU WILL NOT SIT ON THIS MONEY.

I HAVE SUCCEEDED IN CARRYING THE FOUR METAL BOXES OUT OF THE COUNTRY, WITH THE AID
OF SOME TOP GOVERNMENT OFFICIAL, WHO STILL SHOW SYMPATHY TO MY FAMILY, TO A
NEIGHBOURING COUNTRY (ACCRA-GHANA) TO BE PRECISE. I PRAY YOU WOULD HELP US IN
GETTING THIS MONEY TRANSFERRED OVER TO YOUR COUNTRY. EACH OF THESE METAL BOXES
CONTAINS US$5,000,000.00 (FIVE MILLION UNITED STATES DOLLARS ONLY) AND TOGETHER
THESE FOUR BOXES CONTAIN US20,000,000.00(TWENTY MILLION UNITED STATESDOLLARS ONLY).
THIS IS ACTUALLY WHAT WE HAVE MOVED TO GHANA.

THEREFORE, I NEED AN URGENT HELP FROM YOU AS A MAN OF GOD TO HELP GET THIS MONEY IN
ACCRA GHANA TO YOUR COUNTRY. THIS MONEY, AFTER GETTING TO YOUR COUNTRY, WOULD BE
SHARED ACCORDING TO THE PERCENTAGE AGREED BY BOTH OF US.PLEASE NOTE THAT THIS MATTER
IS STRICTLY CONFIDENTIAL AS THE GOVERNMENT WHICH MY LATE HUSBAND WAS PART OF IS
STILL UNDER SURVAILLANCE TO PROBE US.

YOU CAN CONTACT ME THROUGH MY FAMILY LAWYER AS INDICATED ABOVE AND ALSO TO LIAISE
WITH HIM TOWARDS THE EFFECTIVE COMPLETION OF THIS TRANSACTION ON TEL/FAX NO:xxx-x-
xxxxxxx AS HE HAS THE MANDATE OF THE FAMILY TO HANDLE THIS TRANSACTION.

THANKS AND BEST REGARD

MRS. MARIAM ABACHA

FIGURE 13.2
Nigerian e-mail scam letter.

FIGURE 13.3
Fraudulent e-mail link
redirects users to the
scammer's web site.

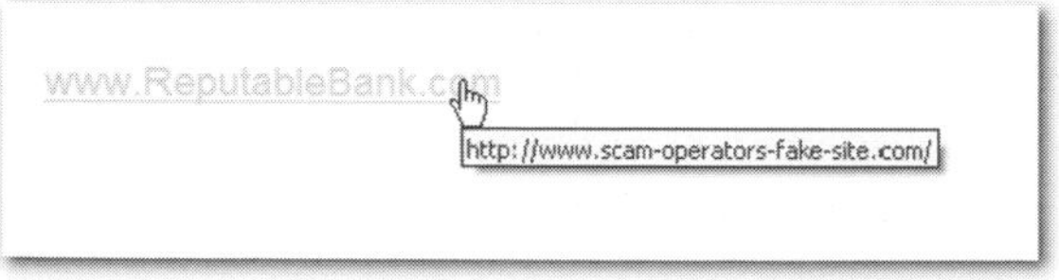

www.ReputableBank.com

http://www.scam-operators-fake-site.com/

Building Trust

As mentioned at the beginning of this section, building customer trust and confidence when doing transactions on the Internet is an important aspect of e-commerce. TNS Research reports on the percentage of shoppers who are concerned about security on the Internet.

To give your web customers a feeling of security, it is wise to become a member of one of the

Percent of shoppers who ...

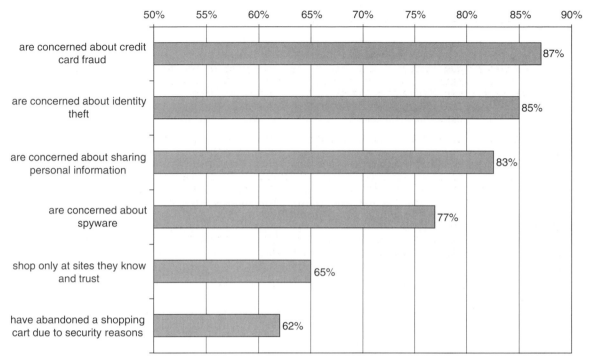

FIGURE 13.4
Online shoppers' security concerns. (Source: TNS Research and Forrester Research, posted at www.verisign.com.)

Internet security verification companies, such as VeriSign or BizRate. Web customers need an easy way to see that their transactions are protected and that they are dealing with reputable firms. In what has emerged as the online equivalent of the Better Business Bureau, these companies provide Internet security and verify the web site owner through a certificate authority. A certificate authority or certification authority (CA) is an entity that issues digital certificates for use by other parties. It is an example of a trusted third party. A September 2007 market share report from SecuritySpace.com determined that the company VeriSign and its acquisitions (which include Thawte and more recently Geotrust) have a 57.6% share of the certificate authority market, followed by Comodo (8.3%), GoDaddy (6.4%), DigiCert (2.8%), Network Solutions (1.3%), and Entrust (1.1%). There are several companies that offer these services at no charge or for a free trial period.

Another alternative is to make your products available with at least one of the reputable online retailers, for those customers who would otherwise hesitate to give out their credit card information to an unknown web site. It may cut in to your bottom line for profit on each unit but will probably increase overall sales and exposure—keeping in mind that 87% of online shoppers are concerned about credit card theft.

Copyright Violation

Copyright is a form of protection of "original works of authorship" so that the creators may benefit financially from their efforts. This includes literary works, music, dramatic performances, artistic creations, and other intellectual works, referred to as *intellectual property* (IP). The original intent of copyright law was to provide this opportunity of exclusivity to the creator but to also guarantee the public right to exploit these works in the public domain after the expiration of the copyright term. The U.S. Constitution states, "to promote the progress of science and the useful arts, by securing for limited times to authors and inventors the exclusive right to their respective writings and discoveries" (U.S. Constitution, Article 1, Section 8).

MUSIC PIRACY

Illegal peer-to-peer file sharing first became big news in 1999 when Shawn Fanning introduced Napster to millions of music fans so that they could easily share recorded music files over the Internet. Since then, the recording industry has taken many steps to protect copyrighted sound recordings, including seeking criminal and civil prosecution of offenders and lobbying governments to enact stricter laws to protect intellectual property.

The Recording Industry Association of America's (RIAA) web site states, "If you make unauthorized copies of copyrighted music recordings, you're stealing. You're breaking the law, and you could be held legally liable for thousands of dollars in damages." Within Title 17, United States Code, Sections 501 and 506 is written the words "Federal law provides severe civil and criminal penalties for the unauthorized reproduction, distribution, rental or digital transmission of copyrighted sound recordings." Making a personal copy of a song that has been purchased is not against the law, but the copy must be for personal use and may not be distributed to another person. Any reproduction beyond this limited use is illegal—even if it is not for financial gain.

When creating a web site that will contain music for sampling, it is very important to include protection for the copyrighted materials. The various options are discussed in Chapter 8.

UNAUTHORIZED USE OF COPYRIGHTED MATERIALS ON THE INTERNET

These days, it's almost too easy to copy and paste items from other web sites onto your own. However, much of the material contained in web sites is subject to copyright protection and thus cannot be used without consent of the copyright holder. According to Virginia Montecino of George Mason University, the following elements of a web page are protected: links, original text (such as articles and blogs), audio and video, graphics, html and other markup languages and software code, and lists that have been compiled by the web site creator or their organization. Among the elements that are legal to use are original works including writings, images, recordings, and videos for which you are the creator or hold the copyright. You can link to other web sites, although you cannot use any trademarked icons, wordmarks, or other protected materials for the link. You can use anything that is specifically designated as free, unrestricted materials, or anything for which you have obtained permission or fulfilled obligations to purchase the rights. You can use limited portions of the works of others, as you would do in a research publication, provided appropriate credit is given to the original source (Montecino, 1996).

Abuse of Personal Information

The web has given marketers the opportunity to learn more about customers than was ever possible before. As a result, consumer advocates are concerned about web users' privacy rights and the abuse of data collected by web sites. There are several ways that Internet users give up personal information while surfing the Web. Each computer connected to the Internet has a unique Internet protocol (IP) address, in the form of four sets of numbers, each set separated by a period. The Internet service provider (ISP) that a person uses knows the IP address of each computer and normally does not share that information. But marketers still have several ways to determine web usage patterns and purchase behavior online. Search engines track activity to help determine how they compile results for search terms. They can record IP addresses for comparison, even though they don't know about the individual conducting the searches. This helps them determine which sites, and which advertisements, are relevant.

The most widely recognized invasion of privacy on the Internet is the use of *cookies*. Cookies are pieces of information stored on a user's hard drive by various web sites to help those sites identify the user when the user returns to the site. Cookies may contain information such as log-in or registration information, preferences, address information, and areas of interest. The site can then use those cookies to customize the display it sends to each user, based on the information in the cookies. However, third-party marketers can use these cookies to profile individuals based on what else they do on the Internet. Typically, these data are used in aggregate to determine the behavior of groups rather than individuals, but this practice has the potential to be abused by targeting individuals for unsolicited advertising messages.

The music service iLike works by monitoring music consumption on one's computer when the individual uses iTunes and then posts that listening information on the user's Facebook page. Ordinarily, that could be problematic for users who would prefer that their friends don't know they still listen to, for example, *Alvin and the Chipmunks* at age 35. iLike has a special feature that allows users to remove these artists. The site states:

> Do you secretly like an artist but don't want the whole world to know? Check off any artists that you no longer want to display in your play data. Note: Newly played artists will be unchecked by default. Artists that are removed will no longer appear in your recent plays or top artists lists and will no longer be counted when calculating your compatibility with others or when the system recommends new music to you.
>
> **(www.iLike.com)**

Spyware, and the related programs adware and malware, are software programs that install themselves on a user's hard drive and gather information as the user travels the Internet. The original use was to collect information for marketing purposes, but it has more malicious uses and often installs itself on the user's computer without their knowledge. It can be used to track keystrokes, revealing sensitive passwords and financial account information. Most antivirus software programs now include antispyware features that will scourer the user's computer, looking for and removing such programs.

Spamming

Spam is defined as unsolicited commercial e-mail messages—the equivalent of junk mail or telemarketers. Because of the negativity and problems associated with spam, successful marketers have adopted a code of conduct. The U.S. government and several foreign countries have passed legislation regulating or outlawing the act of spamming. Many special interest sites (discussion groups) strongly discourage "harvesting" (collecting from posted messages) e-mail addresses to be used for spamming. This is done by gathering up e-mail addresses of people who have posted messages on the site and then using them for sending e-mail messages to them without their permission.

When sending out mass e-mail messages (which should be done sparingly and only to those who have given permission), do not make all the e-mail addresses visible to the other recipients. They may have given you permission to send them e-mails, but that does not extend to one person on your list piggybacking by hitting the "reply all" button and sending out their own marketing message to the addresses on your list.

Spamming can take on other forms besides just e-mail. Bulletin boards are having trouble preventing the bot programs from finding them and bombarding them with spam messages. Spambots are automated programs designed to register on message boards or forums, disseminate the spam messages, and leave. Usually, they leave a fake name and e-mail address and mask their true IP address. These annoying posts include links to commercial websites with the dual purpose of generating traffic to the site and increasing search engine placement. The text may be unrelated to the forum topic, but the increase in incoming links may improve the search ranking for the spammer's sites. In response, some message boards and forums now employ CAPTCHA programs, which require the user to visually identify a string of numbers and letters and then type that in before the message is permitted to be posted (see Chapter 6).

Instant messaging (IM) has become another recent target for spammers, called "spim." The "spimmer" sends out an IM that includes a link in the message. Those who click on the link subject themselves to spyware that could be installed on their computer.

The U.S. Congress, as well as many other governing bodies, has addressed this issue with legislation. A portion of the CAN-SPAM Act of 2003 states:

Whoever, in or affecting interstate or foreign commerce, knowingly—

'(1) accesses a protected computer without authorization, and intentionally initiates the transmission of multiple commercial electronic mail messages from or through such computer,

'(2) uses a protected computer to relay or retransmit multiple commercial electronic mail messages, with the intent to deceive or mislead recipients, or any Internet access service, as to the origin of such messages,

'(3) materially falsifies header information in multiple commercial electronic mail messages and intentionally initiates the transmission of such messages,

'(4) registers, using information that materially falsifies the identity of the actual registrant, for five or more electronic mail accounts or online user accounts or two or more domain names, and intentionally initiates the transmission of multiple commercial electronic mail messages from any combination of such accounts or domain names, or

'(5) falsely represents oneself to be the registrant or the legitimate successor in interest to the registrant of 5 or more Internet Protocol addresses, and intentionally initiates the transmission of multiple commercial electronic mail messages from such addresses,

or conspires to do so, shall be punished as provided in subsection (b).

'(b) PENALTIES—The punishment for an offense under subsection (a) is—

'(1) a fine under this title, imprisonment for not more than 5 years, or both ...

CAN-SPAM Act of 2003

PRIVACY POLICIES

All web sites should post their privacy policies, which should include basics on how the web site operates and protects customer information. The policy should also include the specific steps a person can take to be removed from an e-mail list, and all e-mail correspondence with customers should have an unsubscribe feature. The description should include how and what information is collected from site visitors and how that information may and will not be used. Customers generally do not want their names or e-mail addresses sold to third parties. If you plan to do so, you need a separate opt-out button that allows visitors to sign up for your e-mails but reject those from others, usually referred to on the web site as "our business partners."

The most stringent protection of an individual's right of privacy on the Internet comes from the individual company's privacy policy, generally posted on a web site. In an article for *PC Magazine* titled "What They Know," author Cade Metz wrote:

[T]the biggest limitation on companies' behavior is their own privacy policies, which describe the personal information they collect about you and how that information is used. "Once they pledge to maintain particular standards with respect to user privacy," says Krent [Harold Krent, professor, Chicago-Kent College of Law], "then, according to private law of the contract as well as public law of FTC-type activity or consumer protection at the local level, you can sue for misleading information or failure to abide by the pledges that one makes."

(Metz, *PC Magazine,* **2001)**

Reputable Internet companies all post privacy policies. Those that do e-commerce also post security policies. The policy should offer a way for visitors to contact the webmaster with questions or comments concerning privacy or data security. The following sample privacy policy was taken from Chinook Webs. com (http://www.chinookwebs.com/privacysample.asp).

SAMPLE PRIVACY POLICY

Thank you for visiting our web site. This privacy policy tells you how we use personal information collected at this site. Please read this privacy policy before using the site or submitting any personal information. By using the site, you are accepting the practices described in this privacy policy. These practices may be changed, but any changes will be posted and changes will only apply to activities and information on a going forward, not retroactive basis. You are encouraged to review the privacy policy whenever you visit the site to make sure that you understand how any personal information you provide will be used.

Note: The privacy practices set forth in this privacy policy are for this web site only. If you link to other web sites, please review the privacy policies posted at those sites.

Collection of Information

We collect personally identifiable information, like names, postal addresses, e-mail addresses, etc., when voluntarily submitted by our visitors. The information you provide is used to fulfill you specific request. This information is only used to fulfill your specific request, unless you give us permission to use it in another manner, for example to add you to one of our mailing lists.

Cookie/Tracking Technology

The Site may use cookie and tracking technology depending on the features offered. Cookie and tracking technology are useful for gathering information such as browser type and operating system, tracking the number of visitors to the Site, and understanding how visitors use the Site. Cookies can also help customize the Site for visitors. Personal information cannot be collected via cookies and other tracking technology; however, if you previously provided personally identifiable information, cookies may be tied to such information. Aggregate cookie and tracking information may be shared with third parties.

Distribution of Information

We may share information with governmental agencies or other companies assisting us in fraud prevention or investigation. We may do so when: (1) permitted or required by law; or, (2) trying to protect against or prevent actual or potential fraud or unauthorized transactions; or, (3) investigating fraud which has already taken place. The information is not provided to these companies for marketing purposes.

Commitment to Data Security

Your personally identifiable information is kept secure. Only authorized employees, agents and contractors (who have agreed to keep information secure and confidential) have access to this information. All e-mails and newsletters from this site allow you to opt out of further mailings.

Privacy Contact Information

If you have any questions, concerns, or comments about our privacy policy you may contact us using the information below:

By e-mail:

By Phone:

We reserve the right to make changes to this policy. Any changes to this policy will be posted.

Wolf Leonhardt, Chinook Webs, http://www.chinookwebs.com

Permission granted by Wolf Leonhardt, Chinook Webs (www.chinookwebs.com).

CONCLUSION

The Internet is a new format for marketing and communication and still has that pioneering spirit when it comes to entrepreneurship. Certain mistakes should be avoided and protocol followed to create an atmosphere of professionalism for your customers. The free-for-all aspect of the Internet brings with it the potential for abuse in many forms. Marketers should be aware of these pitfalls so that web sites can accommodate and address the concerns of its visitors.

GLOSSARY

Adware – A form of spyware that collects information about the user in order to display advertisements in the web browser based on the information it collects from the user's browsing patterns.*

CAPTCHA – Short for completely automated public Turing test to tell computers and humans apart, a technique used by a computer to tell if it is interacting with a human or another computer.*

Certificate authority or certification authority (CA) – An entity that issues digital certificates to web site owners to verify the site owner and provide a measure of security to customers. It is an example of a trusted third party assuring customers that the site is legitimate.

Cookies – The name for files that your web browser stores on your hard drive; they hold information about your browsing habits, such as what sites you have visited, which newsgroups you have read, and so on.

Flash – Animated features typically found on high-end web sites and generally created with Adobe software.

Harvesting – Collecting e-mail addresses by visiting user groups and copying e-mail addresses from their message boards.

Instant messaging – A type of communications service that enables you to create a kind of private chat room with another individual to communicate in real time over the Internet, analogous to a telephone conversation but using text-based, not voice-based, communication.*

Intellectual property – "Intellectual property refers to creations of the mind: inventions, literary and artistic works, and symbols, names, images, and designs used in commerce. Intellectual property is divided into two categories: Industrial property, which includes inventions (patents), trademarks, industrial designs, and geographic indications of source; and Copyright, which includes literary and artistic works such as novels, poems and plays, films, musical works, artistic works such as drawings, paintings, photographs and sculptures, and architectural designs." (From the World Intellectual Property Organization, www.wipo.org.)

Internet protocol (IP) address – The particular identification number that each computer connected to the Internet possesses.[3]

Internet service provider (ISP) – A company that provides access to the Internet, generally for a monthly fee.

Payola – A bribe given to a disc jockey to induce him or her to promote a particular record.

Pharming – Similar in nature to e-mail phishing, pharming seeks to obtain personal or private (usually financial related) information through domain spoofing (looks like the real domain but is fake).*

Phishing – The act of sending an e-mail to a user falsely claiming to be an established legitimate enterprise in an attempt to scam the user into surrendering private information that will be used for identity theft.*

Search engine optimization (SEO) – The process of increasing the amount of visitors to a web site by ranking high in the search results of a search engine.*

Secure sockets layer (SSL) – A method of encrypting data as they are transferred between a browser and Internet server. Commonly used to encrypt credit card information for online payments.

---

[3] Depending upon the circumstances of it's connection, the IP address may not be the same every time a computer connects to the Internet. You'll more than likely get the same Class A and Class B numbers (the first two numbers in the IP address), but the other two may be different. It'll also be different if you connect in a different location such as a coffee house or a library.

Spamming – The activity of sending out unsolicited commercial e-mails. The online equivalent of telemarketing or junk mail.

Spimming – Like spamming only targeted using instant messaging instead of e-mail.*

Spyware – Any software that covertly gathers user information through the user's Internet connection without his or her knowledge, usually for advertising purposes.*

Viral marketing – Marketing phenomenon that facilitates and encourages people to pass along a marketing message.

Wordmark – A standardized graphic representation of the name of a company, institution, or product name used for the purposes of identification and branding.

* Definitions from Webopedia (www.webopedia.com).

REFERENCES AND FURTHER READING

Allbusiness.com. Top 10 Internet marketing mistakes, www.allbusiness.com/print/3969-1-22eeq.html.

eConsultancy. Top ten viral marketing mistakes, www.e-consultancy.com/news-blog/362010/top-ten-viral-marketing-mistakes.htm.

Gardner, Jan. (2000). Web site no-no's. Inc.com.

Geisheker, Peter. (2006). Do you make these Internet marketing mistakes? www.marketing-consulting-company.com/mistakes.htm.

Lowery, Shelly. (2006, April 3). 35 Website mistakes that can kill your business. SWeCS Newsletter. www.southwestecommerce.com/newsletters/swecsnews030406.htm.

Lurie, Ian. (2006, April 24). Five big online marketing mistakes. iMedia, www.imediaconnection.com/content/9157.asp.

Maven, Richard. (2006, October). Top ten viral marketing mistakes. E-consultancy, www.econsultancy.com.

Metz, Cade. (2001, November 13). What they know.PCMagOnline. http://www.pcmag.com/article2/0,1759,5562,00.asp.

Montecino, Virginia. (1996). Copyright and the Internet, http://mason.gmu.edu/~montecin/copyright-internet.htm.

Nevue, David. (2005). How to promote your music successfully on the Internet, www.promoteyourmusic.com.

Phang, Alvin. Things you should not do when you market online. Ezine@articles, http://ezinearticles.com.

Pingel, Cheryle. (2004, May 27). Top 10 online marketing mistakes, www.imediaconnection.com/content/3547.asp.

Privacy rights clearinghouse. Privacy and the Internet: Traveling in cyberspace safely. www.privacyrights.org/fs/fs18-cyb.htm.

SecuritySpace.com. (2007, October). Certificate Authority Market Share Report, www.securityspace.com/s_survey/data/man.200709/casurvey.html.

ShoestringBranding. (2007, September 27). The three biggest Internet marketing time wasters. www.shoestringbranding.com/2007/09/27/the-three-biggest-internet-marketing-time-wasters/

Tippins, Robyn. (2005, September 19). What not to do on your website:

Top ten website no-no's. www.associatedcontent.com/article/10129/what_not_to_do_on_your_website.html.

CHAPTER 14
Mobile Media

The digital and mobile revolutions continue to present the entertainment industries with opportunities and challenges. Whereas the Internet and desktop computers have dominated the paradigm shift in the past, the future belongs to wireless mobile technology. Already, the growth of cell phone adoption outstrips Internet adoption. The basic mobile phone, which has been around for nearly 30 years, has evolved from a voice-only form of communication to a multimedia device capable not only of communication but of delivering information and entertainment. In 2008, The International Confederation of Authors and Composers Societies (CISAC) predicted that by 2010, half of all digital content would be delivered to mobile devices. A 2007 survey of 16- to 54-year-old cell phone users by Lodestar Universal found that although only 35% of mobile phone usage in the United States was for features other than voice phone service, over 75% of usage in Japan is for other services.

FROM CELL PHONE TO MOBILE COMMUNICATION DEVICE

The wireless handset, now morphing into a multimedia personal communication device, continues to grow in popularity around the world. Beginning with the third-generation (3G) handsets a few years ago, mobile data networks and increasingly sophisticated handsets have been providing users a variety of offerings. The first generation of mobile phones was introduced in the 1970s, and the period lasted through the 1980s. These devices worked on an analog signal much like those used by two-way radios. The second generation began in the 1990s, used digital voice encoding, and included the geographically compatible Global System for Mobile (GSM). The third generation, or 3G technology, allowed for enhanced multimedia, broad *bandwidth* and high speeds, and usability for a wide variety of communication tasks.

Mike Hanlon (2008) in *GizMag* reported that half of the global population has a cell phone. He stated that we have reached the tipping point in cell phone penetration. According to Hanlon, 1.12 billion phones were sold globally in 2007, up from a billion the previous year. The largest growth is coming from China and India, with many developing companies opting to bypass landlines in favor of setting up the less expensive wireless communication and avoiding the installation of a landline infrastructure.

> Large slabs of Asia and the Pacific have been swept up in the coming of mobile telecoms in the last few years—in January 2004, only one in twenty Pakistanis had a mobile phone. By the end of 2008, better than one in two Pakistanis will own a mobile, taking sophisticated communications into many regions for the first time. The social changes must be immense.

> The Pew Internet and American Life Project reported in early 2008 that over 75% of Americans use either a cell phone or a *personal digital assistant* (PDA). And 62% of all Americans have some experience with mobile access to digital data of some form, whether it is text messaging or music downloading. More Americans said it would be harder to give up their cell phone (51%) than the Internet (45%) or television (43%). In a breakdown by age group, older users were more inclined to say it would be hard to give up their landline telephone than their cell phone—unlike the 18–29 demographic, who would much rather give up a *landline* phone than their cell phone.
>
> **Mike Hanlon, GizMag**

CONSUMER USAGE OF MOBILE COMMUNICATION DEVICES

The 3G handsets, which are now widely in use in parts of the globe, offer users a new data network with more bandwidth than the original analog and first-generation digital cellular phone service. The wireless data services continue to evolve. Marguerite Reardon (2008) wrote in CNET news, "The ink on the checks written to pay for mobile operator's 3G wireless networks is barely dry, and operators are already thinking of their next generation networks" (http://www.news.com/2102-1039_3-6229811.html). The Worldwide Interoperability for Microwave Access, or *WiMAX* platform, offers consumers the ability to utilize all the different types of wireless communication systems in the marketplace; in other words, it offers interoperability between different types of devices including cell phones, PDAs, and laptops. In 2008, Sprint Nextel committed to spending $5 billion on using the technology. The proposed network may cover up to two miles from one base station and

Those who say it would be very hard to give up ...

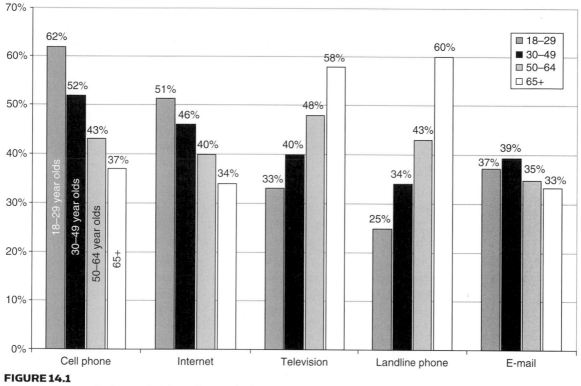

FIGURE 14.1
Consumer willingness to give up entertainment/communication technologies.

deliver speeds of up to 12 Mbps. However, WiMAX faces competition from other standards, such as long term standard (LTE),[1] that are entering the marketplace.

The Pew Internet and American Life Project (2008) has also reported that as of December 2007, 62% of Americans have had some experience with mobile access for activities other than voice communication, either through mobile phones or wireless laptops. The most popular alternate use for cell phones is *text messaging*, with 58% of cell phone users stating they have used their device for text messaging. Equally, 58% have used their cell phones to take a picture, whereas only 17% have used their handheld device to play music and only 10% to watch a video. Of interest to the Pew researchers, nonwhites, specifically African Americans and English-speaking Hispanics, were more likely to engage in media activities other than voice phone service than their white counterparts.

[1] LET is a new wireless standard expected to enable downlink/uplink peak data rates above 100 Mbps/50 Mbps.

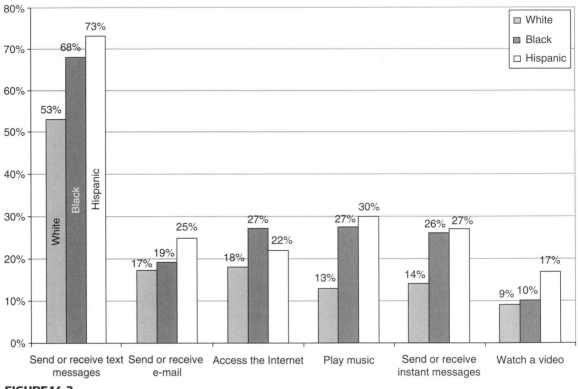

FIGURE 14.2
Mobile communications activities by ethnic background. (Source: Pew Internet and American Life.)

Mobile device usage also varies by age, with younger adults much more likely to use these devices for text messaging and all other nonvoice activities. In all categories, usage decreased with age, thus underscoring the vast generational differences that exist between a generation that has grown up around computers and an older generation whose computer experience is limited to adulthood. In her article, "Text Messaging on Rise with Young People," Martha Irvine (2006) stated, "the former darling of high-tech communication [e-mail] is losing favor to instant and text messaging." The article continued to say that young people see e-mail as a way to reach older people such as their parents, but are, themselves, increasingly using text messaging over e-mail as a means to stay in touch with each other. They associate e-mail with work, school, and spam.

With text messaging already in high gear for younger demographics, and mobile Internet access on the rise in the United States among young people, the United States is expected to catch up to Asia and Europe in Internet use via mobile handsets, offering marketers and musicians a chance to communicate with fans in innovative new ways.

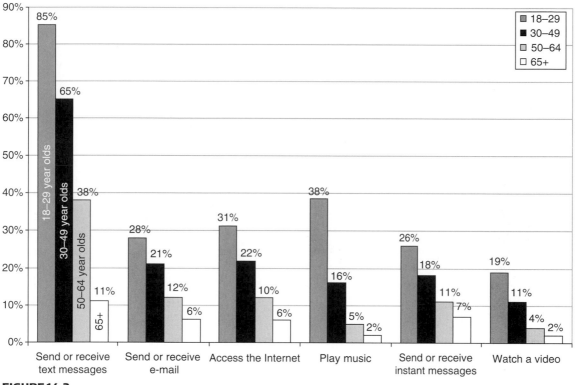

FIGURE 14.3

Use of mobile communications by age group. (Source: Pew Internet and American Life.)

On the m-commerce (mobile commerce) front, in a study by LodeStar Universal—which included the United Sates, Brazil, Russia, Pakistan, India, China, and Japan—e-commerce and financial transactions were touted as showing promise, with 81% of Japanese respondents claiming to have already used the services. In the United States, banks are beginning to offer banking services via cell phone, including checking on account balances, paying bills, and transferring funds, and concert promoters are beginning to use cell phones for ticket sales and delivery (see Chapter 15).

Are We There Yet?

In the article, "The Portable Web Still Has a Long Way to Go," Michael Fitzgerald (2007) lamented the slow pace at which the mobile web business is developing, stating that only 13% of cell phone users in North America use the cell phone to surf the Web more than once a month. He stated that in 2000, the *wireless application protocol* (WAP) was supposed to bring Internet service to cell phone users, but that the adoption has been impeded by lack of high-speed networks and the high cost of services to the consumer. Caroline Gabriel, a Rethink Research analyst, said data made up only 12% of revenue per user in 2007, far below the 50% that was expected. In the "Geek" column (August 2006) in *Internet Exposure LiveWire*, the columnist stated, "three quarters of people are avoiding using mobile Internet because of the poor user experience and the high costs. (Chapter 15 will discuss ways to create a *mobi* web site that is more suitable for small handsets.)

Enid Burns (2008), in the article "U.S. Mobile Web Adoption Slow," stated that "mobile Internet penetration is lower in the U.S. than many European countries." Burns cited that an average of 29% of European Internet users access the Web on mobile devices, compared to 19% in the United States.

Noah Elkin vice president at iCrossing stated, "The cost of mobile Internet access … continues to be a serious impediment for many consumers, and currently it's simply too high for mass adoption" (Elkin, 2007). He also advocated for more web-friendly handheld devices such as the *iPhone*.

Yet by the end of 2007, Sarah Perez, in her article "Mobile Web Use Growing Faster Than Ever," was optimistically stating, "mobile web access is a trend that is growing fast and will continue to grow" (Perez, 2008). She cited a 25% increase in monthly data transfer rate over the course of 2007, adding that the

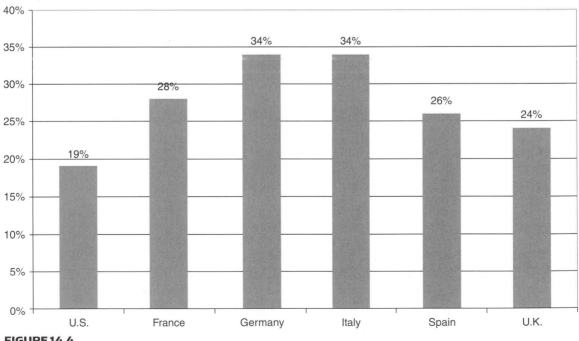

FIGURE 14.4
Mobile Internet penetration by country. (Source: ComScore.)

adoption of more sophisticated mobile devices, along with an improvement in customized content for mobile users, will facilitate further adoption of mobile web use.

The results of the Pew Internet and American Life project indicate that Americans have indeed become dependent on the use of mobile communication devices for activities other than voice phone, especially text messaging. That study and others have shown that mobile data and communication activities are especially important for user groups who do not also have landline Internet service, whether it's in developing countries who have bypassed landlines or demographic groups in the United States and other countries who had opted for wireless communication only and are less like to have ever adopted landline communications. The Pew study also indicates that mobile web adoption is still in the early stages among most demographic groups, with the 18 to 29 age group showing the most usage at 31% of cell phone users. This group is also the most inclined to be heavy users of social networking sites. As mobile Internet becomes more popular, marketers are looking at ways to provide popular content to this age group, and social networking looks very attractive.

MOBILE SOCIAL NETWORKING

In 2006, *Business Week* news analyst Olga Kharif reported that MySpace was creating a partnership with Cingular Wireless to bring MySpace to mobile customers. Already in 2006, 33% of young adults (18- to 24–year-olds) were posting photos to web sites via mobile phones. MMetrics analyst Mark Donovan said in 2006, "there's absolutely interest in participating in mobile social networks" (Karif 2006). Then in 2007, MMetrics reported that 12.3 million consumers in the United States and Western Europe reported accessing a social networking site with their mobile device in the month of June 2007. Kharif stated, "it's the cell phone, rather than the personal computer, that's the constant companion for today's hip and socially networked."

The attraction for marketers is the 3.3 billion cell phone subscribers, which, according to Victoria Shannon of the *New York Times* (2008), is a number that far outweighs total Internet users. Kharif expects mobile customers to use social networking services when they are stuck somewhere outside the home, for example, waiting for public transportation.

Mobile social networks are attempting to exploit the geographic benefits of mobile social networking by providing location information of friends via global positioning information. This added dimension to social networking allows users to physically meet one another using mobile devices. The Web 2.0 site loopt.com touts its service as a social compass by automatically notifying users when their friends are physically nearby. This feature could be beneficial to concert goers who attempt to meet up with friends at a venue that may be filled with thousands of other fans (see Chapter 15 on cell phone usage at concerts).

As mobile phone users continue to find new ways to use phones for staying connected and for social networking functions, opportunities are developed to promote music, much the way computer-based social networking sites have found numerous ways to incorporate music into the social activities of their sites. Chapter 15 will explore ways to use mobile communication devices to promote music and music-related social events with the help of text-messaging, viral marketing, and mobile Internet features.

GLOSSARY

3G technology – The third generation of developments in wireless technology, especially mobile communications, that allows for the transmission of data and multimedia content.

Bandwidth – The amount of data that can be carried from one point to another in a given time period (usually a second).[2]

Global positioning (GPS) – A group of well-spaced satellites that orbit the Earth and make it possible for people with ground receivers to pinpoint their geographic location.

GSM – A digital cellular phone technology that is the predominant system in Europe but also used worldwide. GSM is the dominant second-generation digital mobile phone standard for most of the world. It determines the way in which mobile phones communicate with the land-based network of towers.

Iphone – A smartphone made by Apple that combines an iPod, a touch screen, a digital camera, and a cellular phone. The device includes Internet browsing and networking capabilities. The iPhone supports both WiFi and cellular connectivity, depending on which signal is available.

Landline phone – Refers to standard telephone and data communications systems that use in-ground and telephone pole cables in contrast to wireless cellular and satellite services.

Mobi – Internet domain used for web sites that supply content to cell phones and other handheld devices with tiny screens.

PDA – Personal digital assistant; a handheld device that may combine computing, telephone/fax, Internet, and networking features.

Social networking sites – Places on the Internet where people meet in cyberspace to chat, socialize, debate, and network.

Text messaging – Sending short text messages from a mobile phone to other mobile phone users.

---

[2] Bandwidth only accounts for the time that it takes to get into or out of your connection and is really just a potential speed. Each communication device has it's own bandwidth based upon its transmission capabilities. The speed of any communication is limited by the device(s) that are communicating, sort of a lowest common denominator situation. Bandwidth also doesn't account for periodic bottlenecks anywhere in the system or extra "hops" that the signal may need to make through the Internet.

WAP – Wireless application protocol is a specification for a set of communication protocols to standardize the way that wireless devices can be used for Internet access. A WAP browser provides all of the basic services of a computer-based web browser but is simplified to operate within the limitations of a mobile phone; for example, it has a smaller view screen.

Wi-MAX (Worldwide Interoperability for Microwave Access) – "Designed to extend local Wi-Fi networks across greater distances such as a campus, as well as to provide last mile connectivity to an ISP or other carrier many miles away. In addition, Mobile WiMAX offers a voice and higher-speed data alternative to the cellular networks" (answers.com/topic/wimax).

REFERENCES AND FURTHER READING

Burns, Enid. (2008, October 23). U.S. mobile web adoption slow, www.clickz.com/showPage.html?page=3623758.

Elkin, Noah. (2007, March 29). The path to mobile web adoption. iMedia, www.imediaconnection.com/content/14206.asp.

Fitzgerald, Michael. (2007, November 25). The portable Web still has a long way to go. *International Herald Tribune.* www.iht.com/articles/2007/11/25/technology/proto26.php.

Geek column. (2006, August). Dear Geek column. www.iexposure.com/live_wire/august06/dear-geek.cfm.

Hanlon, Mike. (2008). The tipping point: One in two humans now carries a mobile phone. GizMag. www.gizmag.com/mobile-phone-penetration/8831.

Irvine, Martha. (2006, July 18). Text messaging on rise with young people. www.foxnews.com/wires/2006Jul18/0,4670,SnailEmail,00.html.

Kharif, Olga. (2006, May 31). Social networking goes mobile. *Business Week,* www.businessweek.com/print/technology/content/may2006/tc20060530_170086.htm.

Mi2n. (Music IndustryNews Network). In 2010, half of all digital content will be delivered to mobile devices, www.mi2n.com/press.ph3?press_nb=106654.

Perez, Sarah. (2008, March 11). Mobile web use growing faster than ever. ReadWriteWeb, http://publications.mediapost.com/index.cfm?fuseaction=Articles.showArticle&art_aid=77840.

Pew Internet & American Life Project. (2008, March). Mobile access to data and information, www.pewinternet.org.

Reardon, Marguerite. (2008, February 12). The next big thing in wireless. *CNET News*, www.news.com/2102-1039_3-6229811.html.

Shannon, Victoria. (2008, March 6). Social networking moves to the cell phone. *The New York Times.* http://query.nytimes.com/gst/fullpage.html?res=9C03E1DE143BF935A35750C0A96E9C8B63.

Turakhia, Saurabh. (2007, November 1). The medium is the mobile phone. *Hindustan Times.* www.kiwanja.net/database/article/article_mobile_medium.pdf.

Walsh, Mark. (2008, March 5). Study: Cell phone passes land line as hardest to give up. *Media Post Publications,* http://publications.mediapost.com/index.cfm?fuseaction=Articles.showArticle&art_aid=77840.

CHAPTER 15

Mobile Music Marketing

MUSIC GOES MOBILE

The global mobile music market is expected to rise to more than $17.5 billion by 2012 according to a Jupiter Research study released in early 2008 (Mobile Music Adoption, 2008). This will be driven by subscription music services and full-track downloads. The mobile music market is expected to be more successful in the area of subscription services, unlike the PC-based network, which has seen disappointing numbers for music subscription services. Perhaps the difference is that consumers desire to own music that resides in the household, but they may be more willing to "rent" music on portable devices, much like a personal radio station.

The year 2007 was considered the tipping point for mobile music adoption, according to a Jupiter Research report. Winsor Holden, author of the Jupiter report stated, "far more subscribers began downloading and subscribing to music content in developed markets, and it must be said that the publicity surrounding the iPhone launch undoubtedly contributed to consumer awareness of mobile music services per se" (http://hypebot.typepad.com). Until 2007, the majority of music for mobile devices took the form of *mastertones*—short excerpts from an original sound recording that plays when a phone rings. The market for ringtones, ringtunes, and the like developed throughout the first half of the decade but saw a decline in sales for the first time in 2007 as consumers moved away from phone personalization features and began to adopt full track downloads to mobile (International Federation of Phonographic Industries [IFPI]). Ringtones accounted for 62% of the mobile music market in 2007 but are predicted to account for only 38% of the market by 2012. The decline in ringtones may also be facilitated by consumers creating their own mastertones on their home computers and *sideloading* them into their mobile devices, and also because, in many cases, the ringtones cost more than the whole song.

FIGURE 15.1
Nokia multimedia phone.
(Used by permission.)

While the market for ringtones appears to have peaked in 2007, new opportunities arise for independent artists and labels to feature their music for sale as ringtones. The service Xingtone offers artists the ability to sell their own ringtones to fans through the Xingtone web site. The "silver plan" is the best option for independent artists and allows artists to offer up to 15 different ringtones and set the price on each. The plan costs about $10 per month (www.xingtone.com). MyNūMo provides a similar service, but it does not charge a monthly fee. Instead MyNūMo keeps a larger percentage of the sale price (www.mynumo.com).

MOBILE MUSIC SALES

The IFPI reported that in 2007 "there was a notable pickup in sales of full track downloads to mobile, with sales accounting for 12% of all digital sales in the first half of the year," compared to 6% in the same period a year earlier. There was quite a variation in the popularity of mobile music purchases, with Japan and China leading the way in a comparison of market share between mobile music sales and online sales. The China/Far East region will remain the largest marketplace for mobile music sales, according to Jupiter Research's predictions for 2012. Japan is the first market where sales of full-track mobile downloads, compared to mastertones, are the leading mobile music format, accounting for over 40% of digital sales. The IFPI credits Japan's success on the formation of mobile music retail services jointly owned by the record labels.

In 2007, France also showed a greater propensity for mobile sales than online sales, but that in part is a reflection of the weak online sales market resulting from problems with illegal file sharing and not based on large mobile music sales. France also has a weak digital music market overall, with digital sales accounting for only 7% of the market. The UK, on the other hand, has the most advanced mobile market in Europe with the greatest penetration of music phones—43% of mobile subscribers.

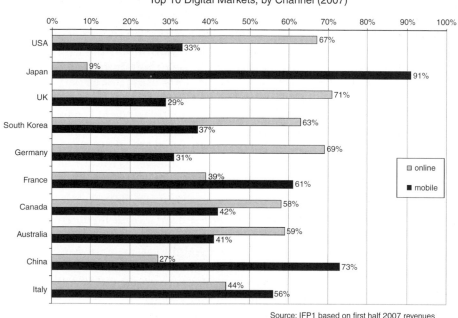

Top 10 Digital Markets, by Channel (2007)

Source: IFP1 based on first half 2007 revenues

FIGURE 15.2
Top 10 digital markets, comparison of online and mobile revenue.

MARKETING

The emerging area of mobile music creates new challenges for marketing music to mobile devices. One of the first questions addressed in marketing is how to deliver the product to the consumer. What form should it take? How should it be distributed? And do the answers apply across the board to all segments of the market? Business models that work on the Internet may not translate seamlessly to mobile devices. The media industries are fraught with failed experiments into new technology where old paradigms were applied without regard to unique features of the new technology. Radio shows were adapted to the new medium of television, but not without drastic modifications along the way. Music has found its way into mobile devices beginning with car radios, through the development of the Walkman, to the portable CD player, the MP3 player, satellite radio, and now the mobile wireless handset.

Rent or Sell?

In the absence of experience in the area of music-to-mobile, questions arise as to whether the consumer will ultimately prefer to (1) purchase and download tracks, much like iTunes; (2) subscribe to a music service much like Rhapsody; or (3) prefer to listen to streaming audio like a personal radio channel, much like Pandora or LastFM. Or perhaps some new business model not yet developed by the industry will become the consumer favorite. In the article "Marketing Mobile Music," John Gauntt lamented, "The most pressing question facing mobile carriers, music labels, and music service providers is how consumer behavior will evolve as mobile music transitions from a phase dominated by early-adopter, active music fans to one more influenced by mainstream, casual music fans." He speculated that the variety of ways that digital music can be packaged and played on a variety of fixed and mobile devices such as dedicated music players, mobile phones, DVD players, stereos, TV, and automotive audio systems make it likely that mobile music will not likely be permanently fixed to a specific device. But even this varies from country to country. According to Gauntt, the U.S. music market leans toward multidevice models, whereas consumers in Korea and Hong Kong prefer listening primarily on mobile phones.

The subscription-based model that has not shown success for Internet music sales does show more potential for mobile music, with consumers more willing to "rent music" that is accessed remotely. Pyramid Research estimates that the global market for "a la carte" mobile music will be worth around $2.5 billion in 2011; with subscription models, mobile music market potential swells to more than $4 billion in 2011 (not including ringtones, etc.).

Subscription services have been a hard sell to Internet users who prefer to possess their music collections on their computers. The subscription services require subscribers to sync their libraries each month to verify their subscription service is still active. This limits consumers' ability to transfer the music to portable devices. Meanwhile, mobile companies and record labels are not enthusiastic about transferring the same iTunes business model to the cell phone because much of the profit from the iTunes model comes from hardware sales, with small profit margins on 99 cent downloads. So according to Colin Gibbs of RCRWirelessNews (2007), "recurring revenue—in the form of subscriptions—may be the key for mobile music services."

Distribution is not a problem and financial transactions are not a problem because mobile users are billed for services monthly. The question becomes how to package and price music for the consumer. Until the industry sorts out what the consumer wants and is willing to pay for, companies that are currently in the digital music sales business, such as iTunes and Napster, wireless carriers such as Verizon and AT&T, and specialty upstarts such as Groove Mobile will continue to experiment with their offerings to consumers until something hits.

Meanwhile, the mobile industry continues to test unique marketing ideas to get music consumers more involved with using mobile devices for music discovery and consumption. Several of those strategies are outlined next.

Music ID

One of the most challenging problems for record labels since the 1950s has been finding ways to help music lovers connect with the music they hear and are then motivated to purchase. A gap existed between listening to the song and determining where and how to find it for sale. FM radio, with its long blocks of music, has done a poor job of helping listeners identify the music they are hearing, and therefore, the label loses out on a sale. A campaign in the late 1980s attempted to encourage radio stations to back-announce songs they played with a reminder to disk jockeys: "when you play it, say it." In the 1990s, early versions of retail kiosks sought to provide reference information that would help consumers quickly locate music they were searching for. Some kiosks provided music samples, others had a keyword index. The Internet introduced other methods for identifying music, the most important being the music sample, which allowed consumers to recognize and confirm they had found the titles they intended to purchase, either online or in the store.

One of the more recent features of mobile phones is a service that helps users identify music so that it can be purchased. Verizon offers V-Cast Song ID. When users hear a song being played on a loudspeaker, over a radio, and so forth, they can press a few buttons to call up the service and hold the phone near the music source. The service will then identify the song and send a text message back to the user with the song title and artist. AT&T does the same thing with its MusicID feature. Verizon also offers the user the option to download the song from its music store directly to the user's multimedia phone.

Advertising via Cell Phones

Although the majority of advertising on cell phones takes place through *text messaging*, marketers are beginning to look at other options. In a study of teens in the UK (Q Research), respondents said they preferred to receive advertisements in the form of picture ads, with text ads being the least popular. In the study,

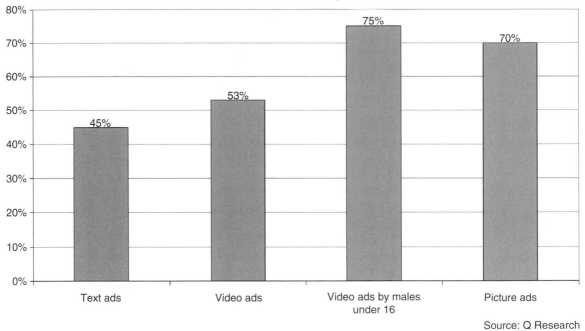

Type of advertising UK teens would prefer to receive on their cell phones

Source: Q Research

FIGURE 15.3
Type of advertising UK teens would prefer to receive on cell phones.

only males under 16 preferred video ads, with other respondents expressing concern over the cost of receiving video ads.

Marketing company Echo Music developed a campaign for an artist that included sending an audio advertising of country artist Dierks Bentley's voice telling fans about the release of a new album and playing a music sample from the album. Bentley's voice kicked off the message stating, "Hey, dudes. Today's my big day. I'm excited for ya'll to hear it," followed by a snippet from the album.

Push/Pull and Viral Marketing via Text Messaging

Cell phone marketers have begun using push messages to consumers' cell phones to market products. A push message is a specially formatted *short message service* (SMS) text message that gives you the option of connecting directly to a particular wireless application protocol (WAP) (mobile Internet) site, using your phone's browser. In other words, the text message has a link; by clicking on the link, the user's cell phone is directed to a special mobile-friendly web site. According to Nester and Lyall (2003), "The majority of mobile campaigns today use (SMS) messaging. SMS messages are limited to text only and to 160 characters. Despite the limitations of the media, it has proven to be highly effective generating high brand recall and response rates."

Pull marketing campaigns are those that target the customer to take some form of action, such as requesting or demanding a product or visiting a web site or store to learn more about a product. For cell phone use, a pull strategy may use some other form of media to encourage consumers to use their cell phones to look up a web site or respond to a challenge. Nester and Lyall used the example of a "text to win" campaign, or special offers that encourage users to log on using their cell phones to receive free wallpaper or ringtones. The TV show *American Idol* uses a pull strategy on the live program to encourage viewers to text in their vote using their AT&T phones.

Push campaigns, on the other hand, rely on the marketer sending push messages to the cell phones of potential customers. This requires that consumers "opt in" to allow the marketer to send these messages. Even then, fear of, or rejection of, spam messages may cause consumers to opt out of the process unless there is some incentive to grant permission. Marketers use enticements to encourage permission. A study of teens in the UK revealed that only 32% of teens are willing to receive ads on their mobile phones, with no enticements. These teens prefer to receive picture ads on their phones, rather than videos or text only. When asked if they would be willing if they only received push ads on things they are interested in, the positive response jumps to 71%. And, when incentives are added in, the response is even higher. In fact, in late 2007, one cellular provider, Blyk, launched a phone service in the UK that is free to consumers—it is fully advertiser funded. The service targets 16- to 24-year olds, by offering 217 texts and 43 minutes of voice calls every month for free in return for users agreeing to receive six daily push marketing messages.

The results of this study indicate that push ads need to be tightly targeted to the receiver's interests. Nester and Lyall reported that push campaigns have a response rate of about 12%, much higher than other forms of direct marketing.

Viral marketing campaigns are new to cell phones and incorporate the notion of recipients passing along the message to others. In the blog "How to make viral marketing appealing for phones," JayVee (2006) mentioned three factors to successful viral campaigns on cell phones:

1. Offer exclusive content: something that is not available on the Internet but only via cell phones. It could include ringtones, free music files, wallpaper, graphics, and so on.
2. Make it useful and timely: for musicians, something centered around a live concert date would be effective, offering something of value related to the concert or the artist.
3. Clearly define objectives: the goals should be simple and straightforward; create awareness for the product or service; encourage recipients to pass along the message; elicit a call to action (log on, download, etc.).

There are several Web 2.0 sites that facilitate group SMS messaging and are available to end users, making it perfect for the do-it-yourself (DIY) music marketer. GupShup is a site that allows users to easily start an SMS group. Musicians can

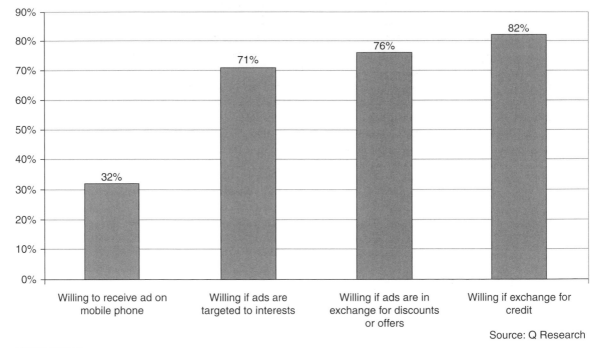

Percent of UK teens willing to accept mobile ads

Source: Q Research

FIGURE 15.4

Percentage of UK teens willing to accept mobile ads.

start a group, invite friends and fans, and have a convenient method to send out blast messages in support of their career. The GupShup site outlines three simple steps to start a GupShup from your mobile phone:

1. Create your group by sending an SMS to 567673434. Example: CREATE mybandname.
2. Invite friends to your group. SMS the following to 567673434: INVITE fan1, fan2, fan3, friend1, friend2, friend3.
3. SMS, your message to 567673434 to post a message to your group. Example: Hi, we're the Rock Dudes and this is our group of friends and fans. Please check out our new recordings at ….

The service allows you to send group text messages to mobile phones from your phone, encouraging them to pass along the message and respond to the call for action. New groups can issue up to 50 free invitations or up to 200 opt-in invitations. You then can send out a mobile text SMS message to all members of the group, such as *2night @ bluesvl lounge ROCK DUDES @ 8 dont B L8*.

Another service, TextHQ allows you to "send text messages from your PC to mobile phones. Send text messages to one phone or group of phones instantly. Upload pictures to your gallery and send them to mobile phones worldwide."

TextHQ is not a fee service, but prices are reasonable at 250 messages for under $30, and the service offers delivery notification. Zlango is a platform that transforms SMS into a colorful icon-based experience; a combination of text and icons can be sent to mobile phones. Their plug-ins and widgets allow for some creative communication, but as of April 2008 it was not available in the United States. ZipToPhone is a free text messaging service. Using this simple tool, you can, from your computer browser, easily send SMS messages to a cell phone. Mozes is a service that boasts, "Take your fan club mobile … Broadcast text messages to your various groups and get quick replies, or poll them and see instant results on the web too. You can ask your mobile mob trivia questions, send coupons, run contests and more." Kadionk offers to mobilize your social networking (www.kadoink.com). Twitter has features and widgets to interface between the web and mobile phones at www.twitter.com.

SMS MESSAGING SERVICES

GupShup: www.smsgupshup.com

TextHQ: www.texthq.com/index.asp

Zlango: http://zlango.com/products

ZipToPhone: http://ziptophone.com

Mozes: www.mozes.com

Kadoink: www.kadoink.com

Twitter: www.twitter.com

More links at www.go2web20.net.

Proximity Marketing

Proximity marketing is the localized wireless distribution of advertising content associated with a particular location, generally a retail establishment, transportation hub, or venue (Wikipedia). The electronic marketing messages are delivered to willing customers' handsets and may include store coupons and discounts. A press release from Ace Marketing and Promotions states:

> What if you could tap every potential customer on the shoulder within 100 m of your location and send them a business card, advertisement or special offer directly to their mobile phone? What if you could do this constantly, without worry about the cost?

Proximity marketing allows businesses to do this as long as the customer has the equipment to receive and has agreed to accept marketing messages. Messages should not be sent to customers who have not opted-in to receive them. Bluetooth-enabled cell phones and PDAs, as well as cell phones equipped with a GPS system, are among the current generation of mobile devices that are capable of being targeted by proximity marketing. Proximity marketing services are capable of sending text messages, video clips, ringtones, audio files, and promotional or discount coupons complete with a graphic representation of bar codes. An article in South Africa's BizCommunity.com touts eight advantages of proximity marketing over other forms of marketing (from Proximity Marketing Revealed, Imajinn):

- *Speed.* Content is delivered to mobile devices in seconds
- *Cost.* Significantly cheaper than direct mail or other media
- *Relevance.* Content to users in relevant places at relevant times
- *Interactivity.* Interact with or engage consumers nearby
- *Proximity.* The mobile device is always in the consumer's pocket
- *Reach.* 100% of current mobile phone market, for example
- *Personalization.* Bespoke content per individual consumer
- *Viral.* Content shareability factor allows for peer-to-peer transfer

Proximity marketing is also capable of responding to pull marketing, where consumers seek out nearby retail establishments by logging on to the Internet using their mobile devices. The system would note the location and then send a customized list of appropriate retailers in the vicinity of the consumer's device. For example, someone traveling who would like to know the names and locations of nearby pizza parlors could simply log on, type in "pizza," and get a local list.

One of the most relevant applications of proximity marketing in the music business is proximity marketing at concerts and venues. One of the tenants of proximity marketing is having members of the target market in the area who will be responsive to the marketing messages. In all applications of proximity marketing, these messages need to have some appeal to the recipient in the form of (1) information they may desire to have, (2) special offers, (3) privileged access or opportunities, or (4) "advertainment," something that entertains as well as advertises.

Concerts

Marketers have been finding new ways to incorporate cell phone marketing into live performances since the mass emergence of cell phones at concerts began to influence shows in the 1990s. Although recording devices and traditional cameras are not allowed into concert venues, use of cell phones is actually encouraged lately and incorporated into the concert experience. Journalist Neil Strauss of the *New York Times* wrote in 1998 of the emergence of the use of cell phones at concerts for something other than talking: fans were holding up cell phones, allowing someone not in attendance to hear the music. This first display of cell phone interactivity at concerts was just the beginning. A few years later, bloggers started to write of their experiences at concerts where the crowd was prompted by the artist to hold up their opened clam-shell phones to bask the audience in the blue glow from cell phone screens. At that point, the cell phone had replaced the old (and hazardous) tradition of holding up lit cigarette lighters at concerts. Now, some acts will provide specific wallpaper for fans to download that they can use when they participate in the "screen glow" events. In an effort to engage music fans in interactive marketing endeavors, many new forms of interaction via the cell phone have emerged to enhance the concert experience that maximize marketing potential for the artist. The advantage for artists is that they generate a list of phone numbers to be used for future marketing projects. Generally fans are given an incentive to register their cell phone numbers, even if it's just a screen saver, wallpaper, or a music track.

Text Messaging

Major concerts are now routinely offering display screens above the stage that allow for fans to use their cell phones to post a text message on the screen, for a fee or in exchange for registering their phone number and agreeing to accept push messages in the future. For smaller venues such as clubs and parties, a service called partyStrands allows attendees to post their text messages on video monitors to request songs, answer trivia questions, vote on drink specials, or ask someone to dance. The event coordinator only needs to hook up a computer with Internet access to a video projection system. The service derives its revenue from advertisers who pay to have their messages pop up or roll across the bottoms of the screen (www.partystrands.com).

Some major acts have incorporated text messaging into other aspects of their concerts. The band Korn encouraged fans at the concert to vote by phone on which song should end their show. Others have featured contests in which concert goers enter via their cell phone for a chance to win better seats, VIP passes, or other prizes. Fans who are encouraged to participate in these concert-related activities submit their phone numbers and grant the artist permission to send them future marketing messages about the artist. As fans leave the show, the artist has the ability to send them each a message thanking them for coming to the show and offering discounts on merchandise (LaPlant, 2008).

Ticketing

Cell phones are beginning to replace paper tickets in a variety of settings and are expected to soon replace the paper tickets for events and air travel. In addition to reducing the environmental impact and costs, it provides a convenience to both the concert promoter and the ticket holder. Fans can purchase digital tickets on their cell phones, and then a graphic with the bar code is sent to the ticket holder. Fans simply have to scan the onscreen bar code at the entry gate to gain access to the concert. The concert promoter then has access to phone numbers of all ticket holders, many of whom may have opted in to receive future marketing messages.

Concert Music Downloads

As music in the United States moves toward mobile handsets, another opportunity is created to sell recorded music to fans—impulse purchases of concert recordings. Still in its infancy at the time of this writing, there is great potential for artists to "sell back" the concert experience to attendees, even before they've reached their cars in the parking lot. Portions of the event can be audio recorded, mixed on the fly, and offered to attendees for sale via cell phone as they leave the venue. Many times concert attendees have lamented that they wish they had a recording of a concert they just attended, and in the future, they will have the opportunity to purchase one on the spot.

PREPARING YOUR WEB SITE FOR MOBI

Cell phone access to the Internet is bringing new challenges to web design. In the article "Web Sites Adapt to Mobile Access," Amanda Kooser described some of those challenges: the small screens, the different formatting of the various mobile devices, limited maneuverability, and so on. Mobile handsets rely on the *wireless application protocol* (WAP) to create web sites specifically formatted for mobile devices. The domain system used for wireless devices is the *.mobi* (top level) domain extension on the end of the URL address.

Just as with an Internet web site, design for a mobi site must be preceded with an understanding of goals for the site. What are you trying to accomplish and provide with a mobile web site? The goals may not be the same as for the computer-based site. An Internet site might focus on more detail, more long-term information, more entertainment, and creating branding and customer awareness. A mobile site will most likely be used by fans on the go who need quick access to information that they can use immediately. This may include information on live shows, such as directions to venues and start times. A tour schedule may be important if groups of fans are together and want to attend an event at a future date. A quick glance at your tour schedule may be what they are looking for. Contests and coupons may be important for fans as well as access to music to download and listen to. Pull messages that appear on billboards, bus boards, or in newspapers may create an immediate impulse among viewers to respond to the message via cell phone, such as signing up to win something or downloading new content. With these goals in mind, a separate web site must be developed that allows mobile fans to quickly access what they are looking for without scrolling through a lot of pages and screens. And the site must be updated often to reflect the changing needs of its visitors.

General WAP Formatting Rules

Creating mobile web sites may be as simple as stripping out all CSS formatting to reveal a text-only site similar to how a search engine spider renders a site, although much of the content must also be reduced for easier browsing. Or better yet, professional "mobi" developers can create a site that dynamically produces formatting from a variety of options depending on the type of device accessing the site. In other words, the formatting of the site can vary, and for each visitor, the presentation is customized after the server detects what type of handset the visitor is currently using.

Here are some basic rules to consider:

- Be realistic about what will fit on the small screen of a mobile device. Paring down your information for the screen size is first and foremost.
- Eliminate all information that is not important to the mobile user. Less is more on a mobile site. Visitors are usually looking for specific information or completing a specific task.
- Avoid vertical scrolling. Web users don't like it, and it's even less popular with mobile users.

- Reduce the number of clicks. Don't go deep in page numbers; it's slow going for the mobile user.
- Keep it clean. Use readable text on a readable background.
- Test your design on a variety of handheld devices to see if it holds up on all of them.
- Use abbreviations and succinct wording wherever possible
- Access keys are helpful for speeding up navigation. (Access keys are those just below the screen on the left and right that can take on a variety of functions, usually with the name of that function on the bottom of the screen just above the key.)

Mobile Site Design Programs

Mobile web sites are still in the infancy stage of development, with new design ideas entering the marketplace almost every week. As of early 2008, the most common design elements include a small logo and company or site name at the top of the screen, followed by a short menu of options.

There are several professional web developers, such as TrioVisions, creating mobi web sites for major corporations and top-selling musical acts. Among the do-it-yourself options, Dev.mobi touts itself as mobile development community. Its development software is powered by mobiSiteGalore. The mobiSiteGalore offers a free WYSIWYG mobile web site builder including multiple pages, a link manager, and an image editor. Go Daddy offers .mobi web designer tools with its WebSite Tonight feature. Wirenode offers web design tools in a free, easy-to-use WYSIWYG format that allows for text, links, graphics, analytics, and some minor formatting. Jagango offers free mobi sites, a series of templates and dialog boxes for easy formatting, and a mobi community for networking.

For artists who want to be ahead of the pack, creating a "mobi" site now may increase marketing success with a small investment and may generate incremental sales and broaden the fan base. The combination of viral/push text messaging and

FIGURE 15.5
Nokia phone with Yahoo! browser screen.

having a mobile web presence allows for two-stage marketing campaigns, where a push message is sent to fans, who then can respond to the call for action by visiting the mobile site to buy music, check the tour schedule, or enter contests.

MOBI DESIGN WEB SITES

Wirenode: www.wirenode.comJagango:www.jagango.com

WebSite Tonight: www.godaddy.com/gdshop/hosting/hosting_build_website. asp?ci=9028

Dev.mobi design site: http://dev.mobi

Mobi Site Galore: www.mobisitegalore.com.

Updated information at www.wm4mb.com.

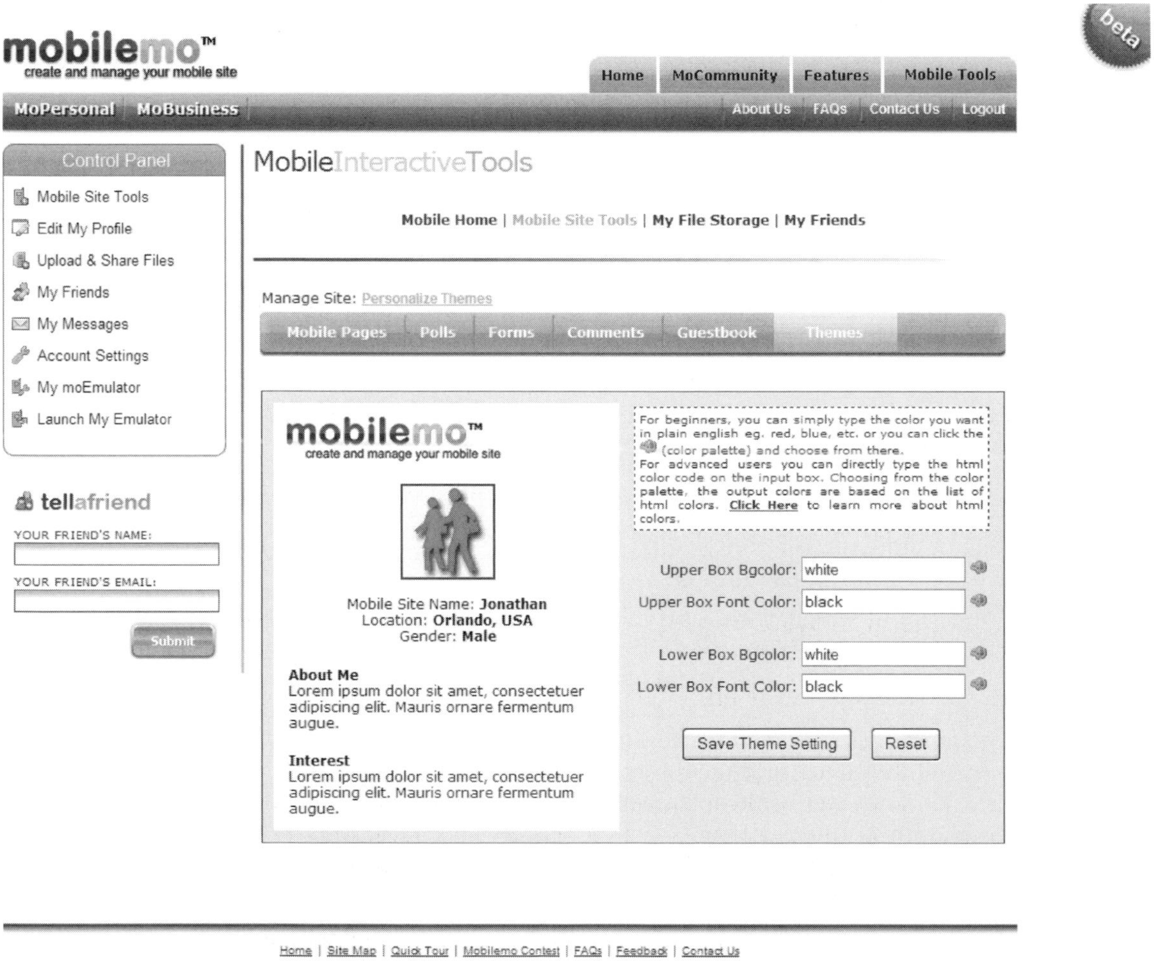

FIGURE 15.6

Setting up a mobi web site. (Courtesy of www.mobilemo.com.)

GLOSSARY

CSS – Short for cascading style sheets, a feature added to HTML that gives both web site developers and users more control over how pages are displayed. With CSS, designers and users can create style sheets that define how different elements, such as headers and links, appear. These style sheets can then be applied to any web page.

Mastertones – Short excerpts from an original sound recording that plays when a phone rings.

Mobi – Top-level Internet domain used for web sites that supply content to cell phones and other handheld devices with tiny screens. The .mobi suffix was introduced in 2005.

Mobile music – Music that is downloaded to mobile phones and played by mobile phones. Although any phones play music as ringtones, true "music phones" generally allow users to import audio files from their PCs or download them wirelessly from a content provider.

Phone personalization features – Customized phone features such as ringtones, wallpaper, and skins that distinguish a person's cell phone as unique and allows consumers to express themselves.

Push message – A specially formatted cell phone text message that gives you the option of connecting directly to a particular mobile Internet site, using your phone's browser.

Short message service (SMS) – Usually refers to wireless alphanumeric text messages sent to a cell phone.

Sideload – A term used in Internet culture, similar to upload and download; the process of moving data between two electronic devices without involving the local computer in the process.

Text messaging – Also called SMS (short message service); allows short text messages to be sent and received on a mobile phone. Messages can be sent from one phone to another by addressing the message to the recipient's phone number.

Text to screen – A system allowing event participants to send text messages that are displayed on a screen at the event.

Top level domain – The last part of an Internet domain name—that is, the letters that follow the final dot of any domain name.

XML (eXtensible Markup Language) – A standard that forms the basis for most modern markup languages. XML is an extremely flexible format that only defines "ground rules" for other languages that define a format for structured data designed to be interpreted by software on devices. XML by itself is not a data format (phonescoop.com).

WAP – Wireless application protocol; a protocol designed for advanced wireless devices; allows the easy transmission of data signals, particularly Internet content, to microbrowsers built into the device's software.

REFERENCES AND FURTHER READING

Alderman, John. (2004, March 11). Mobile phones are the new lighters. Textually. org, www.textually.org/textually/archives/2004/03/003233.htm.

Gartner, Inc. Gartner says consumer spending on mobile music will surpass US$32 billion by 2010. Press release, www.gartner.com/it/page.jsp?id=500295.

Gauntt, John. (2006, May 26). Marketing mobile music, EMarketer, www.imediaconnection.com/content/9747.asp.

Gibbs, Colin. (2007, October 24). Carriers continue struggles with music models. RCR, Wireless News, www.rcrnews.com/apps/pbcs. dll/article?AID=/20071024/FREE/71024009/1066.

Hall, Bennett. (2007, January 29). Digital technology of party Strands links real and virtual world partying. *USA Today*. www.usatoday.com/tech/news/techinnovations/2007-01-29-party-strands_ x.htm.

Holton, Kate. (2007, September 24). Ad-funded mobile operator launches. Reuters News Service. www.reuters.com/articlePrint?articleId=UKL2454798720070924.

Houghton, Bruce. (2008, February 28). Mobile music on $17.5B track despite ringtone decline. *Hypebot*, http://hypebot.typepad.com/hypebot/2008/02/mobile-music-on.html.

Imajinn. (2006, August 25). Proximity marketing revealed. BizCommunity.com, www.bizcommunity.com/Article/196/16/11399.html.

International Federation of Phonographic Industries. Digital Music Report. www.ifpi.org.

Jayvee. (2006, November 28). How to make viral marketing appealing for phones, www.cellphone9.com/how-to-make-viral-marketing-appealing-for-phones.

Kharif, Olga. (2007, December 13). Mobile ads: Not so fast. *Business Week*, www.businessweek.com/technology/content/dec2007/tc20071212_903658. htm.

LaPlant, Alice. (2008). MP3 unplugged: rethinking the digital music future. Information Week, www.informationweek.com/shared/pritableArticleSrc.jhml:jsessionid= DTLSCSY ...

Leeds, Jeff. (2007, August 15). They've just got to get a message to you. *New York Times*. www.nytimes.com/2007/08/15/arts/music/15conc.html?ex=134483 0400&en=72b66706de55f76e&ei=5090&partner=rssuserland&emc=rss.

Mobile music adoption revenues set to reach $17.5bn by 2012. (2008, February 27). http://wirelessfederation.com/news/category/mobile-music/

Mobile phones are the new lighters. (2006, January 24). Stereogum Blog. http://stereogum.com/archives/cell-phones-are-the-new-lighters_002256.html.

Nester Katharine, Kurt and Lyall. (2003). Mobile Marketing: A Primer Report. FirstPartner Research and Marketing. www.a2.com/docs/mobile-marketing.pdf.

Pearce, James. (2003, November 7). Picture this: Tickets on your cell phone. CNET News, www.news.com/2101_1026_3-5103968.html?tag=st.util.print.

Soeder, John. (2007, May 24). Cell phones becoming the ticket at concerts. *The Plain Dealer*. Cleveland, OH, www.vibes.com/corporate/news/2007/052407Cleveland-Concerts.html.

Strauss, Neil. (1998, December 9). *New York Times*, http://query.nytimes.com/gst/fullpage.html?res=9C01E2DF153AF93AA357 51C1A96E958260.

Wikipedia. (2008). Proximity marketing.

CHAPTER 5

Graphic Editing Software

GIMP for Windows www.gimp.org
Serif PhotoPlus www.freeserifsoftware.com/software/PhotoPlus/default.asp
Paint.net www.getpaint.net
Pixia http://park18.wakwak.com/~pixia/download.htm
ImageForge www.cursorarts.com/ca_imffw.html
Ultimate Paint www.ultimatepaint.com
XnView www.download.com/XnView/3000-2192_4-10067391.
    html?tag=lst-0-5
Saint Paint Studio www.download.com/Saint-Paint-Studio/
    3000-2192_4-10066321.html?tag=fd_sptlt

For more photo editor software, try the following:
www.photo-freeware.net
www.download.com
www.softpedia.com

Web Development Software

Microsoft Expression Web
    www.microsoft.com/expression/products/overview.aspx?key=web
Front Page http://office.microsoft.com/en-us/frontpage/default.aspx
Adobe Dreamweaver www.adobe.com/products/dreamweaver
CoffeeCup www.coffeecup.com/software
Nvu http://nvudev.com/index.php
Namo WebEditor www.namo.com
WebStudio www.webstudio.com
SiteSpinner www.virtualmechanics.com
Web Easy www.v-com.com/product/Web_Easy_Pro_Home.html
Web Builder www.wysiwygwebbuilder.com
Web Plus www.freeserifsoftware.com/software/WebPlus/default.asp
Updated content available at www.WM4MB.com.

Templates

Template Monster www.templatemonster.com
4Templates www.4templates.com
FreeWebsiteTemplates www.freewebsitetemplates.com
Generate your own dummy text at
    www.webdevtips.co.uk/webdevtips/codegen/clickgo.shtml

CHAPTER 6
Resources for Scripts

Online HTML Code Generator http://htmlcode.discoveryvip.com
Click & Go drop down list generator
    www.webdevtips.co.uk/webdevtips/codegen/clickgo.shtml
Meta tag and SEO generators www.submitcorner.com
Guestbook and mailing list script generators www.javascript.nu/cgi4free
HTML and CSS scripts www.hypergurl.com/htmlscripts.html
Updated information available at www.WM4MB.com.

Check Your Web Site

Check your web page HTML at http://validator.w3.org or
    www.anybrowser.com/validateit.html
Check your page's/site's CSS at http://jigsaw.w3.org/css-validator
Check your links at http://validator.w3.org/checklink or
    www.anybrowser.com/linkchecker.html
Check your images for accessibility issues at
    http://juicystudio.com/services/image.php.
Check your content for readability at
    http://juicystudio.com/services/readability.php
Check to see if your CSS's text and background colors have sufficient
    contrast at www.accesskeys.org/tools/color-contrast.html
Check how your site is viewed on different systems at
    www.anybrowser.com/siteviewer.html, www.anybrowser.com/
    ScreenSizeTest.html www.browsercam.com/Default2.aspx
Check your page's performance and web page speed at
    www.websiteoptimization.com/services/analyz
Check several things at once with
    www.netmechanic.com/products/HTML_Toolbox_FreeSample.shtml

CHAPTER 7
Keyword and SEO Tools

Wordtracker keyword suggestion tool:
    www.wordtracker.com
Google AdWords keyword tool:
    https://adwords.google.com/select/KeywordToolExternal
Digital Point Management keyword tool:
    www.digitalpoint.com/tools/suggestion
SEO Tools: www.seochat.com/seo-tools/keyword-suggestions-google
Keyword verification tool: www.marketleap.com/verify

Tell a Friend Scripts

www.plus2net.com/php_tutorial/tell_friend.php
www.javascriptkit.com/script/script2/tellafriend.shtml

Statistics and Web Analytics Tools

www.OneStatFree.com provides a free, simple stat counter service and more advanced fee-based services.

www.Alexa.com monitors web traffic of people who use its toolbar. The site provides lists of most popular sites.

www.trafficestimate.com is a free tool to help estimate the volume of traffic that a web site gets.

www.statcounter.com is a service that offers both free and paid statistics service.

www.iwebtrack.com is another service with free and paid versions of the service.

http://Analytics.google.com (see above)

Updated information available at www.WM4MB.com.

CHAPTER 8
Just a Few of the Many Music Players for Your Web Site

PremiumBeat free MP3 player: www.premiumbeat.com

The Wimpy Player: www.wimpyplayer.com

Secure-TS Player: fancy skinned secure flash player for under $30. www.tsplayer.com

XSPF Music Player: www.musicplayer.sourceforge.net

Video Editing Tools

Riva FLV encoder: (free trial) www.rivavx.com/?encoder

ConvertTube: www.converttube.com

Flvix: a simple free online video converter. www.flvix.com

Gaffitti: online video editor. http://graffiti.vidavee.com

FlowPlayer: http://sourceforge.net/projects/flowplayer

Tutorial on how to add video to a web site: www.boutell.com/newfaq/creating/video.html

Tutorial on how to add video to a web site: www.2createawebsite.com/enhance/adding-video.html

Flash player: www.macromedia.com/software/flash/about

CHAPTER 9
Major Online Retailers

iTunes www.itunes.com

CD Baby www.cdbaby.com

Amazon.com www.amazon.com

Snocap www.snocap.com

The Orchard www.theorchard.com

C/Net's download.com http://music.download.com

Amie Street http://amiestreet.com/page/for-artists

Tunecore (www.tunecore.com), SongCast (www.songcastmusic.com), and
IRIS (www.irisdistribution.com).
Sell swag through Zazzle at www.zazzle.com or Café Press
www.cafepress.com/cp/info/sell

CHAPTER 10
Music Industry Trade Associations

National Association for Recording Merchandisers (NARM)
www.narm.com
National Academy of Recording Arts and Sciences (NARAS)
www.grammy.com
American Association of Independent Music www.a2im.org
International Federation of Phonographic Industries (IFPI) www.ifpi.org
Songwriters Guild of America (SGA) www.songwriters.org
American Federation of Musicians www.afm.org
Audio Engineering Society www.aes.org
National Association of Recording Industry Professionals (NARIP)
www.narip.com
National Association of Broadcasters www.nab.org
International Alliance for Women in Music www.IAWM.org

EXAMPLES OF INDUSTRY CONFERENCES

Billboard Magazine sponsored events
www.billboardevents.com/billboardevents/index.jsp
South by Southwest SXSW Music & Media Conference: Austin, TX.
www.sxsw.com
Midem: Palais de Festivals, Cannes, France. www.midem.com
Radio and Records Magazine R&R Convention
www.radioandrecords.com/Conventions/RRconvention.asp
Millennium Music Conference http://musicconference.net/mmc12
Audio Engineering Society www.aes.org/events
National Association of Recording Merchandisers
www.narm.com/esam/AM/Template.cfm?Section=Convention

CHAPTER 11
Resources for E-zines and Distribution of Press Releases

The Ultimate Band List www.ubl.com
CD Baby www.cdbaby.org
BeatWire www.beatwire.com
The Ezine Directory www.ezine-dir.com
Music Industry News Network www.mi2n.com
MusicDish www.musicdish.com (try the open review and "submit your
article")

John Labovitz's E-ZINE LIST
www.e-zine-list.com/titles_by_keyword/music/page1.shtml
Zinester Ezine Directory www.zinester.com/Music-
Dmusic http://news.dmusic.com/submit
PRWeb www.prweb.com
Updated information available at www.WM4MB.com.

Resources for Internet Radio Promotion

Rhapsody: www.Rhapsody.com
SomaFM: http://somafm.com
Pandora: www.pandora.com
Live365: www.live365.com
SHOUTcast: www.shoutcast.com
Digitally Imported: www.di.fm
International Webcasting Association: www.webcasters.org
Radio and Internet Newsletter: www.kurthanson.com
New! www.clearchannelmusic.com/cc-common/artist_submission/index.
htm?gen=2

Resources for Artists to Facilitate Their Online Presence

Provides artists with an online profile page and widgets:
www.reverbnation.com
A site that features electronic press kits: www.sonicbids.com
A place to feature your music, videos and photos: www.Virb.com
Provides artists with an online profile page and widgets: www.ripple9.com
Encourages fans to pass along artist information: www.BuzzNet.com
Updated information available at www.WM4MB.com.

CHAPTER 12

Lists of Social Networking Sites

The Social Network list:
http://mashable.com/2007/10/23/social-networking-god
The Wikipedia list:
http://en.wikipedia.org/wiki/List_of_social_networking_websites
Author's updated list: www.internetmarketingforthemusicbusiness.com
The ultimate web 2.0 reference site: www.go2web20.net

CHAPTER 15

Ringtone Sales Services

Xingtone www.xingtone.com
MyNūMo www.mynumo.com

SMS Messaging Services

GupShup www.smsgupshup.com
TextHQ www.texthq.com/index.asp
Zlango http://zlango.com/products
ZipToPhone http://ziptophone.com
- Twitter: www.twitter.com
- Kadoink: www.kadoink.com

Mobi Design Web Sites

Wirenode www.wirenode.com
Jagango www.jagango.com
WebSite Tonight
    https://www.godaddy.com/gdshop/hosting/hosting_build_website.
    asp?ci=9028
Dev.mobi design site http://dev.mobi
Mobi Site Galore www.mobisitegalore.com
Updated information at www.wm4mb.com.

Index